Uncertain Climes

Uncertain Climes

DEBATING CLIMATE CHANGE
IN GILDED AGE AMERICA

Joseph Giacomelli

The University of Chicago Press   *Chicago and London*

The University of Chicago Press, Chicago 60637
The University of Chicago Press, Ltd., London
© 2023 by The University of Chicago
Published 2023
Printed in the United States of America

32 31 30 29 28 27 26 25 24 23   1 2 3 4 5

ISBN-13: 978-0-226-82443-7 (cloth)
ISBN-13: 978-0-226-82444-4 (e-book)
DOI: https://doi.org/10.7208/chicago/9780226824444.001.0001

Library of Congress Cataloging-in-Publication Data

Names: Giacomelli, Joseph, author.
Title: Uncertain climes : debating climate change in gilded age
America / Joseph Giacomelli.
Description: Chicago : The University of Chicago Press, 2023. |
Includes bibliographical references and index.
Identifiers: LCCN 2022026641 | ISBN 9780226824437 (cloth) |
ISBN 9780226824444 (ebook)
Subjects: LCSH: Climatic changes—United States—Forecasting—
History—19th century. | Climatic changes—Effect of human beings
on—United States—History—19th century. | United States—
Climate—Forecasting—History—19th century.
Classification: LCC QC903.2.U6 G53 2023 | DDC 363.700973—dc23/
eng/20220816
LC record available at https://lccn.loc.gov/2022026641

♾ This paper meets the requirements of ANSI/NISO Z39.48-1992
(Permanence of Paper).

I promise nothing complete;
because any human thing supposed to be complete,
must for that very reason infallibly be faulty.

HERMAN MELVILLE,
Moby-Dick

CONTENTS

Questions "Forever Remain"

In his 1873 report on Golden Gate Park, engineer William Hammond Hall suggested that "heavy plantations" and other park features would ameliorate San Francisco's climate. "Certain kinds of trees," Hall wrote, are "effective in removing moisture, by precipitation, from the lower stratum of the air." His report argued that a pattern of alternating groves and glades would check excessive winds, induce favorable amounts of rainfall, and regulate humidity levels. According to Hall, park-induced climate change would go hand in hand with improvements in the health and morals of the population. Indeed, Hall's theories of anthropogenic climate change were part of a grandiose vision for environmental change.[1]

Unlike some of his contemporaries, however, Hall did not adhere to a straightforward ideology of landscape improvement through human intervention.[2] Hall viewed society's influence as minute compared with the scale and mystery of nature. In his view, humans were to the environment as "the microscopic polyp to the coral formations of the ocean." But humanity still functioned as the "great disturber of natural laws" and often exerted a negative influence on the "pristine balance of physical nature." Hall argued that human-induced climatic changes would help restore the environmental equilibrium that had been disturbed by deforestation, swamp drainage, and other aspects of nineteenth-century expansion and industrialization.[3]

Hall's proto-ecological views on restoration were rooted in contemporary climate science. Like his thoughts about nature, the engineer's scientific vision reflected a mixture of humility and grandiosity: Hall proclaimed that modern systems of "extended observation and systematic arrangement" had "supplied the data for great advance in meteorological science." Yet he also emphasized that issues related to "man's ability to moderate climate" would "perhaps forever remain questions for further investigation and continued dispute." Hall believed that "natural laws" existed but that many of them would remain uncertain and unknowable. He accepted the existence of scientific uncertainty and incorporated it into his plan for Golden Gate Park.[4]

The Realm of the Unknown

Euro-Americans and others had been debating possible climatic changes long before Hall issued his report. Across the globe, settler colonialism and intensive resource extraction spurred discussion about whether humans could modify climate, sometimes for the better and sometimes for the worse.[5] Colonial projects promised to unlock the latent climatic potential of arid and semiarid environments that purportedly had been misused by indigenous people.[6] In formerly forested regions, clear-cutting and erosion prompted fears about desiccation and increasingly erratic weather patterns. Although Americans had been questioning society's role in shaping the climate for centuries, starting about 1870, increasing deforestation—coupled with the prospect of widespread Euro-American settlement in the Great Plains and Intermountain West—galvanized the debate.[7]

Hall's piece was an early salvo in a long and contentious climate change debate that stretched from the years after the Civil War into the Progressive Era. Like Hall, most writers emphasized the possibility of local and regional climatic changes. Others, meanwhile, perceived small-scale changes as evidence of continental or even global climatic mutability.[8] Hall and many others believed that humans could alter climates through afforestation, deforestation, agriculture, irrigation, reclamation, and a variety of other actions. A competing group of thinkers sought to refute the notion that society could modify climatic conditions. Across the United States, newspapers, government reports, and scientific journals "teemed" with arguments and counterarguments about climatic changes.[9] Some of these purported changes were already under way, and others loomed as future opportunities or catastrophes. Many authors used climate change theories in a struggle over the future of two intertwined projects: westward expansion and capitalist development.

When surveying the cacophonous range of Gilded Age climate writings, it is tempting to divide climate thinkers into two opposing camps: hardcore expansionists who believed in human-induced climate change versus their opponents, a group that advocated for a more measured approach to settlement and development by questioning notions of climate improvement. It might also be tempting to characterize climate change theories as the last vestiges of a folkloric tradition, as impressionistic notions destined to be supplanted by increasingly professionalized and institutionalized disciplines such as climatology and meteorology. The often-cited Gilded Age belief that "rain follows the plow" has prompted generations of scholars to dismiss late nineteenth-century climate change writings as a "footnote to American intellectual history" testifying to expansionist hubris.[10] Especially in the Great Plains, a cadre of writers proffered dubious climate improvement theses. Booster-scientists including Charles Dana Wilber, Samuel Aughey, and Richard Smith Elliott used climate theory as part of their efforts to attract settlers to Kansas and Nebraska. Their writings evinced confidence in Euro-American agricultural settlement in semiarid regions as well as a seemingly boundless faith in humanity's ability to transform and improve landscapes and climate. Along with their allies in Washington, DC, such as the surveyor-scientist Ferdinand V. Hayden, these Great Plains climate theorists have attracted substantial scholarly and popular attention.[11] Elliott, Wilber, Hayden and other seemingly venal authors make for a convenient contrast with more deliberate and methodical figures such as John Wesley Powell, Grove Karl Gilbert, and other writers who criticized facile climate improvement theses as well as heedless expansionism. The long-standing interpretation of nineteenth-century climate theory, with unscrupulous boosters arrayed against principled professional scientists, encourages acceptance of a simplistic morality tale—a parable about the dangers of bad science.

Yet Hall and many climate theorists do not easily fit into either camp. On closer examination, the climate politics of the Gilded Age appear as chaotic and fractious as the broader cultural politics of the late nineteenth century. Even seemingly skeptical scientists such as William Ferrell acknowledged the possibility of some forms of human-induced climate change. Other climate theorists, like forester Bernhard Fernow, challenged the hypothesis that landscape changes could increase rainfall levels but tentatively supported the theory that forest cover could cause precipitation to be more evenly distributed and less violent.[12] Paradox and uncertainty permeated the writings produced over the course of the climate debate. Many authors acknowledged that the causes of beneficial or deleterious climatic changes remained in the realm of the unknown. Even ardent expansionists sometimes admitted that capitalist development had damaged landscapes and climates.[13] And sometimes determined advocates of human-induced climate change such as Hall admitted the insignificance of humanity before the scale and mystery of nature.

The scientists, boosters, and surveyors who argued about climate change in the late nineteenth century struggled with, embraced, and used uncertainty in myriad ways. The term "uncertainty" is not a neologism in the context of climate debates: some nineteenth-century climate thinkers relished pointing out the uncertainty inherent to knowledge production. The famous conservationist George Perkins Marsh dedicated a section of his 1864 tome *Man and Nature* to the "uncertainty of our meteorological knowledge," and in 1883 Henry Allen Hazen described "the great uncertainty arising" from climate data.[14] But clearly not all late nineteenth-century climate theorists expressed their ambivalence about society's ability to understand and improve climate. Several writers legitimized the dispossession of Native Americans by invoking Euro-American settlers' certain and singular power to transform climatic conditions. Many Gilded Age climate writings undoubtedly reflect a belief in certain, positivistic science as well as a hubristic attitude toward American expansionism.[15] As Hall's report demonstrates, however, certain proclamations were often closely intertwined with more uncertain environmental and scientific rhetoric. The imbrication of certain and uncertain attitudes toward science, climate, and environment reveals the limits of the dichotomies often used to frame the cultural and climatic politics of the Gilded Age: climate change believers versus their skeptical opponents, conservationists versus rapacious boosters, and modern science versus naturalistic premodern beliefs.

For the people involved in Gilded Age climate debates, uncertainty loomed as either an intractable obstacle or a source of power and legitimacy. Uncertainties spanned a range of interrelated categories: scientific, cultural, environmental, epistemological, ontological, psychological, collective, individual. Some authors viewed uncertainty as a driving force on the path toward scientific discovery. Others used uncertainty as a powerful political instrument. In the context of climate debates, many writers expressed indeterminacy in spatial and temporal terms: they argued that humans could modify climate only on certain scales and that scientific consensus would remain elusive only as long as data sets were incomplete.[16] A few climate theorists discussed more fundamental uncertainties, acknowledging the limits of human ability to know and understand complex phenomena. Amid the contentious cultural and scientific politics of the late nineteenth-century United States, many climate theorists thrived not by denying mysteries and unknowns, but rather by recognizing, coping with, and communicating uncertainty.

Uncertain Science

Late nineteenth-century writers produced an astounding volume of articles and reports about climate change. Whether in eminent institutions like the Smithsonian or in small-town agricultural society meetings, scientists, surveyors, academics, farmers, journalists, and many others presented their theories about human-induced climate modification. These people backed their claims with many forms of evidence. Individual and collective climate memories often intermingled with empirical, quantitative data. Despite the persistence of anecdotal evidence, Gilded Age climate thinkers often derived their theories from a burgeoning collection of climate data. In the decades after the Civil War, the US Army Signal Corps, the Smithsonian Institution, state weather bureaus, land grant universities, newspapers, boosters, and independent weather aficionados observed, recorded, and cataloged annual rainfall totals and other climate information.[17]

This array of data gave rise to a chaotic multiplicity of climate change hypotheses. Many theorists focused on the influence that forest culture and agriculture had on climate patterns. Some, like Winslow Watson of New York, asserted that the removal of forests near Lake Champlain had caused desiccation.[18] Others, such as Hall and a group of Great Plains boosters, argued that desiccation or aridity could be mitigated or reversed with tree plantations or agriculture. Still others denied both theories, using a variety of arguments. In 1867, for example, the climate theorist John Disturnell argued that surface conditions on the Great Plains were a consequence, not a cause, of climate patterns. Businessman Theodore C. Henry, on the other hand, depicted climate as immutable in a series of addresses he delivered in Kansas in 1882.[19] Theories in favor of climate change were even more eclectic; climate change proponents disagreed about the exact mechanisms through which humans could modify atmospheric conditions. Whereas forester Franklin B. Hough argued that natural and artificial forests initiated a "cooling process" that induced passing clouds to "send down filaments of rain," agricultural boosters like Samuel Aughey emphasized that newly plowed soil could evaporate humidity into the air, creating dew and precipitation.[20] Theories rooted in forest culture and agriculture represented only one component of the vast panoply of Gilded Age climate change theses. Fires, artificial ponds and reservoirs, reclaimed swamps, electrical currents, railroads, and other factors shared the stage with trees and crops as potential influences on climate.[21] And not all writers attributed climatic changes to specific causal mechanisms, with some invoking "nature," divine intercession, or unknown and unknowable causes.

The confusing profusion of climate theories testifies to the uncertainty of scientific knowledge production. Instead of consolidating climate beliefs and stabilizing scientific practice, the data collection efforts of the nineteenth century contributed to a culture of uncertainty, prompting explorer, geologist, and ethnographer John Wesley Powell to complain about the state of climate science. In 1892 Powell wrote an article about flooding and the possibility of anthropogenic climate change. After asserting that the "mighty powers" that determine how "clouds gather and dance in aerial revelry" are beyond the reach of humans, Powell allowed that society might be able to influence climate patterns to a limited extent. Powell believed that climate change theories contained a "modicum of truth" but had been appropriated by a variety of interests and distorted beyond recognition. Scientific knowledge production had thus become too fluid, unpredictable, and prone to reinterpretation. According to Powell, uncertainty was something to be avoided, and scientists should resist the urge to delve too deeply into the realm of the unknown.[22] But not all agreed with Powell. Some of his contemporaries acknowledged or even embraced the paradox at the core of nineteenth-century climate science: that the modern drive to understand and control natural forces created new scientific and cultural uncertainties.[23]

Defining "climate" and "climate change" presented another challenge for Powell and like-minded contemporaries. In the present, "climate" generally refers to long-term phenomena while "weather" refers to short-term events. But "climate," in the words of historian James Rodger Fleming, is more than just the "the average condition of the atmosphere compiled from weather statistics." For Fleming, climate is "inseparable from the temporality and specificity of the social world." Similarly, geographer Mike Hulme writes of climate as an "idea that takes shape in cultures and can therefore be shaped by cultures."[24] Gilded Age climate writers sometimes used the terms "climate" and "weather" (or "climatology" and "meteorology") interchangeably. The expansive notion of "climate" connoted precipitation, temperature, wind, and other phenomena as well as a series of ever-changing relationships—between land and air, between trees and moisture, or between human bodies and their immediate surroundings.[25] The complex, shifting meaning of "climate" added another layer of uncertainty to debates about the temporal and spatial scale of any possible climate changes. W. H. Larrabee, for instance, remarked on the uncertain meaning of "climate" in 1892: "We are embarrassed when we undertake to define climate and what marks to accept as its characteristics."[26]

After the turn of the century, meteorologists increasingly accepted uncertainty and probability as necessary components of daily and weekly weather forecasting.[27] Despite the frequent slippage between "climate" and "weather," meteorologists privileged short-term data and predictions. Powell and other theorists interested in longer-term climatic changes had to contend with years' and decades' worth of confusing and sometimes conflicting data. Possible human-induced climate changes presented such a quandary that they prompted climate researchers to theorize uncertainty even during the late nineteenth century.

In 1888, for example, George Curtis argued that new data-collecting instruments had not eliminated any of the uncertainty surrounding climate change questions. Curtis was a prominent climate theorist affiliated with the Smithsonian Institution and the Department of Agriculture. His 1888 article on potential variations in rainfall in the "Trans-Mississippi" region explained that "it would seem that the . . . data would easily furnish the means for giving a decisive answer to such questions; yet it has been found that, owing to the changes in observers, instruments, exposures and methods of observing, much uncertainty inheres in the results."[28] Curtis implied that standardizing observers and instruments might eventually fulfill the promise of modern science to answer climate change questions. Yet five years later Curtis expressed a deeper skepticism about the ability of statistics to resolve climatological quandaries. His article on the "causes of rainfall" asserted that since "statistical explanations" do not provide an "explanation of the process by which a change in the rainfall may be brought about, they have not helped clarify the misty meteorological conceptions which are current thereon."[29] Despite his willingness to engage with uncertainty, Curtis still seemed frustrated by his inability to answer climate change questions.

His contemporary Gustavus Hinrichs, by contrast, savored the seeming inevitability of scientific uncertainty. Something of an iconoclast, Hinrichs resented the Washington-based scientific bureaucracy to which Curtis and Powell belonged. He established the independent Iowa Weather Service in the 1870s and studied a variety of weather and climate phenomena, ranging from tornadoes and derechos to climate change, during his time as a professor at the University of Iowa.[30] Using seasonal and yearly rainfall data from his network of observers in Iowa, Hinrichs identified an "intimate relation between the percentage of surface covered with timber and the distribution of rainfall."[31] Although Hinrichs found a correlation between afforestation, deforestation, and rainfall patterns, he believed that the causes and mechanisms behind climate change remained mysterious. In order to finally resolve questions about the nature of climate change, Hinrichs envisioned the creation of a centuries-long climate observation network complete with "self-registering instruments." Hinrichs was a firm believer in statistics—he derived a logarithm that determined the relative agricultural usefulness of a rainfall event—and in the existence of natural laws—he claimed the discovery of "every new law . . . endows us with new powers over nature."[32] Hinrichs's writings reveal that belief in fixed natural laws did not preclude belief in the mysterious, and perhaps unknowable, nature of climate and climate change.

Even as he advocated for an almost high-modern system of data collection that would discover nature's laws, Hinrichs retained a sense of humility about society's ability to grasp environmental phenomena. He admitted that his hypothetical observation network would need to be in place for centuries and still might not be able to furnish definite conclusions about climate change.[33] For Hinrichs, uncertainty was not a cause for frustration, but a source of relief. He seemed to share philosopher-psychologist William James's tendency to equate uncertainty with possibility, potential, and hope.[34] In a practical sense, continued uncertainties necessitated the "proper continuation" of Hinrichs's work at the Iowa Weather Service. In a more abstract sense, constant questions and uncertainties represented a more useful means of stepping "into realms of the unknown" than any "unwarranted recrudescence of Baconian empiricism."[35] It is not entirely clear which scientific paradigm Hinrichs might have advanced as an alternative approach to older, seemingly dogmatic ways of knowing such as "Baconian empiricism" or what historians of science Lorraine Daston and Peter Galison describe as "truth-to-nature" approaches.[36] But Hinrichs seemed to believe that embracing complexity and uncertainty would allow scientists to better engage with the mysteries of climate. As I will discuss in chapter 3, Hinrichs was in some ways a singular and idiosyncratic character. But many of his contemporaries, including a group of Illinois climate theorists, shared his fondness for uncertainty.

At their annual meeting in December 1871, members of the Illinois State Horticultural Society engaged in a lively debate about climate modification. Jonathan Periam presented a paper linking "terrific conflagrations"—presumably the Peshtigo Fire, among other great blazes of that year—to "fortuitous circumstances incident to meteorological phenomena occasioned by man himself, in clearing up the country." In his lecture on meteorology, J. H. Tice introduced the theme of uncertainty. Tice sought to extricate climate science from the "tangled fens of materialism" and railed against "the vicious [a]ssumption that the Universe is a piece of mechanism." William Baker echoed Tice by arguing that uncertainty and mystery made meteorology a singular and "attractive" subject: "It is the more so from the very little knowledge we have of it, and from the boundless field it opens for theorizing." At the same time, however, Baker complained about climate theorists who "enjoy mounting a steed which they fancy Pegasus, and careening off through the dim mists of cloudland." The Illinois climate theorists reflect the tensions and dialectics that marked the climate debate. Periam made his argument in a confident tone, depicting the great fires of 1871 as retribution for humanity's environmental and climatic sins. His scientific certainty betrayed a cultural uncertainty about the costs of American industrialization and urbanization. Baker, meanwhile, embraced uncertainty as the driving force behind science but also worried that it made meteorology prone to dilettantism and amateurish flights of fancy.[37]

Other climate theorists, such as Joseph Lovewell, undertook a more radical embrace of uncertainty than Baker. Lovewell served as Kansas state meteorologist from 1885 to 1895 and taught a variety of scientific subjects at Washburn College in Topeka.[38] He portrayed human-induced climate change as an unsettled question "well worthy of all the attention" it was receiving. To some extent, Lovewell shared Baker's paradoxical ambivalence: he speculated about "forces in nature yet undreamed of" without abandoning his belief in objective scientific "truth."[39] Lovewell differed from Baker in that he challenged distinctions between science and other types of knowledge. He appreciated the participatory and communal nature of data collection as well as popular climate theories and folklore. In contrast to some climate theorists who blamed Native Americans for damaging the climate by setting fires and cutting down trees, Lovewell admitted that "the Kaws and Pottawatomies who once traversed these prairies were probably wiser in this kind of weather-lore than the present denizens of our State."[40] Lovewell advocated for a more uncertain, democratic science characterized by popular engagement and a more inclusive attitude toward knowledge production. In an 1882 article, Lovewell's fellow Kansan H. K. McConnell articulated a vision for climate science in which "theories are no longer authoritatively announced. . . . They are given to the public only to elicit and promote intelligent and fair criticism."[41] McConnell and Lovewell viewed citizen science and public engagement as crucial for coping with vexing issues such as climatic changes.

McConnell and Lovewell seemed to believe their uncertainty would not detract from their legitimacy as participants in the climate change debate. Like the Kansans' work, the climate-related writings of Bernhard Fernow, one of the founders of American forestry, illustrate the power conferred by uncertainty. Fernow wrote extensively about the relation between society, forests, and climate. In an 1892 report laden with ecological language, he described "the interdependence between vegetations and meteorological, soil, and water conditions."[42] According to Fernow, the complex, mutually influencing relationship among humans, landscape, and climate made it difficult for scientists to identify the exact causal mechanism at the root of any potential climatic changes. Fernow prefaced many of his statements with qualifications about the uncertainty surrounding climatic questions.[43] Yet his belief in the complexity and uncertainty of climate agencies did not prevent Fernow from making confident policy proclamations. "Whatever the truth," wrote Fernow, "and neither the claimants nor the objectors in forest climatic influences have brought incontrovertible proof," the stakes in the climate debate were simply too important for inaction. The probable but still uncertain role of forests as "needful regulators and preservers of climatic and hydrological conditions" necessitated "keeping certain areas under forest cover" through activist public management of forests.[44] Instead of shrinking from uncertainty and complexity, Fernow incorporated them into a powerful forestry and climate change platform.

Uncertain Natures

Fernow's work on ecology and climate illustrates the links between the development of modern climate science and Gilded Age debates about "nature." Climate theorists constantly invoked nature. When Thomas P. Roberts wrote in 1885 that "arguments can be produced from nature to support any theory," he testified both to the ubiquity of scientific uncertainty and to nature's prominent role in the climate debate.[45] Like the meaning of climate data, the meaning of nature in the context of climate change discourse remained contested and uncertain.

For H. R. Hilton, evidence provided by nature clashed with evidence provided by climate statistics. In 1880 Hilton delivered an address in Topeka titled "Effects of Civilization on the Climate and Rain Supply of Kansas." His speech described a quandary: in spite of overwhelming anecdotal evidence of anthropogenic climate change, data in Kansas showed little or no statistical increase in rainfall. In Hilton's words, "the records of man"—in this case the Signal Service—"and those of nature seem to be in conflict." Hilton sided with the latter, promising to "take only such proofs as nature itself affords" and implying that the evidence of nature, though uncertain, would eventually trump statistics.[46] Some climate theorists shared Hilton's view of nature as a benevolent force while others depicted nature either as inscrutable and indomitable or as a domain destined to be conquered by society. Gilded Age climate thinkers' portrayals of nature reflect a paradoxical mixture of humility and hubris.

Surprisingly, the same people who believed in the enduring mystery of the causes of climate change sometimes advocated complete human domination of nature. J. H. Tice, the Illinois expert who acknowledged the limits of certainty in climate science, envisioned a utopian future marked by the control of nature. At an 1870 horticulturists' meeting in Galesburg, Illinois, one of Tice's colleagues described efforts to "ameliorate our climate" by planting trees as "feeble endeavors" that paled in comparison with vast natural forces over which "human beings have no control." Tice responded by reminding the other horticulturists of the potential for human progress. He argued that society could "[unlock] the mysteries of nature's economy in all her departments" and "not only modify climate, but all the operations of nature."[47]

In contrast to Tice, Lovewell implied that uncertain science could go hand in hand with a belief in the insignificance of humanity relative to nature. Lovewell's 1892 article "Human Agency in Changing or Modifying Climate" stressed that "all the combined power of man is as nothing when brought into collision with mighty forces of nature." Lovewell sought to reconcile his humble view of nature with his belief in anthropogenic climate modification by considering a range of spatial and temporal scales. He argued that "changes of climate occur in the order of nature" while invoking both "human history" and the "geological records" to prove that "changes have occurred in terrestrial climate."[48] In addition to assessing possibilities of climate change on a geological and global scale, Lovewell also explored microclimatology, arguing that cities influenced local climates.[49] Although Lovewell viewed the "mighty forces" of nature as potential collaborators in climate modification, his caution and uncertainty prevented him from endorsing grand initiatives to transform the climate of the American West.

More strident climate change proponents, such as surveyor Cyrus Thomas, depicted nature as a steadfast ally in civilization's quest to convert the Great Plains into a lush agricultural utopia. In a US Geological Survey report, Thomas asked settlers to follow "the plan nature herself has pointed out" and settle the Plains from west to east, starting on the Front Range in Colorado and heading east along the rivers and streams.[50] The anthropomorphized "nature" of Thomas and other climate change proponents often remained uncertain even as it collaborated with Euro-American settlers. In 1878 the *Dodge City Times* described a mysterious intercession by nature in the climate of the Plains: "The advancing waves of settlement roll over a country, and are driven back; there seems to be, for a few years, a line dividing the humid from the arid region, and beyond it no settlement or cultivation is possible. Suddenly it vanishes. No one can tell exactly where; no one knows how; no one can explain why."[51]

Whereas the *Dodge City Times* described a miraculous transformation in the climate of the Great Plains, some climate thinkers articulated visions of environmental degradation and catastrophe. Eastern writers such as Winslow Watson and Hiram Adolphus Cutting followed in G. P. Marsh's footsteps by linking settlement and landscape modification to climate changes such as increased droughts and violent storms. Cutting described the dire consequences of deforestation-induced climate change in Vermont. Using a local-scale approach, he explained how clear-cutting had altered "what we might call paths for showers," leading to the drying up of "formerly perennial" springs and streams. Like Fernow, Cutting warned that the climate changes were still uncertain but too serious to be ignored: "We may call these changes local if we will, and believe their effect on atmospheric conditions will be as difficult to determine in the future as in the past; but let these changes go on for a generation or two longer as they are now going on, becoming general instead of local, and who doubts their effect on both climate and rainfall, as well as water-supply?"[52] Though it may be tempting to separate authors like Cutting from western climate theorists, eastern protoconservationists and Great Plains writers drew from the same intellectual tradition. Easterners were not the only ones to characterize nature as vulnerable. Even Kansas and Nebraska—the heartland of "rain follows the plow" hubris—produced myriad theories of climatic degradation.

Kansan Frederic Hawn, for instance, devised a pan-western theory of desiccation. Hawn believed that precipitation in the Great Plains originated when mountain snows evaporated directly into the atmosphere. He advanced this hypothesis in a series of articles culminating in an 1881 piece titled "The Source of Rains in Kansas." After citing some experiments conducted in the Sierra Nevada, Hawn concluded that denuding mountains of forests would expose snowfields to more sunlight, shorten the evaporation period, and thus disrupt the "close climatic relations that exist between the mountains and Kansas." Hawn feared that these "radical changes" would have severe consequences for Kansas agriculture.[53] It is hard to imagine a starker contrast than that between Hawn's warning and the rosier vision presented by the *Dodge City Times* a few years earlier. Yet Hawn and the newspapermen both believed in some form of anthropogenic climate change, and both advocated westward expansion. The disparity between these two views underscores the tension between anxiety and hope that marked the advent of modern American environmental thought.[54]

Hawn's climate writings also reveal the interconnections between uncertainties about science or nature and the broader cultural politics of Gilded Age America. Professor Legate, the researcher who inspired Hawn's Sierra Nevada-based climate theory, endorsed railroads as potential environmental saviors while lamenting the disastrous climate change caused by expansion and deforestation. Legate argued that constructing new railroads toward Oregon would create "another broad belt of denudation, the influence of which will be to . . . partly restore that equable temperature that formerly prevailed."[55] Nineteenth-century climate theory was rife with restoration rhetoric.[56] Even Cutting, the Vermonter who assailed timber cutters seeking "immediate profit," held that that the moneymaking potential of reforestation projects would persuade Americans to replant clear-cut areas and thereby restore disturbed climates.[57] Climate theorists' faith in restoration highlights the nineteenth-century dialectic between devastation and renewal, a process that continually transformed catastrophes into "lucrative opportunities for redevelopment."[58]

In contrast to Legate and Cutting, however, some authors articulated more fundamental critiques of both the railroad and the plow—another symbol of American progress and expansion.[59] Iowan J. L. Budd, for instance, argued that plowing the Great Plains, along with the "accompanying drainage of sloughs, soils and streams" had damaged the climate by robbing it of moisture. Some of his contemporaries, such as Wilber and Aughey, viewed the plow as an almost sacred instrument of climate improvement. Many agricultural boosters exalted the climatic benefits of both tree planting and plowing. Budd's rebuke of widespread plowing is proof of a schism among believers in human-induced climate change.[60] It also serves as evidence of the deep cultural uncertainties that accompanied scientific doubts and questions.

Reclaiming Uncertainty

The pervasiveness of uncertainty in late nineteenth-century climate discourse poses several intractable questions: How can we evaluate pronouncements of scientific and cultural uncertainty? Did Gilded Age climate theorists truly believe that facts and truths were ephemeral? Or did they strategically produce uncertainty to cast doubt on their opponents and to secure more funding, support, and legitimacy? To what extent were expressions of cultural uncertainty merely submerged under a tide of expansionist rhetoric? The persistence of these questions highlights yet another paradox: that studying uncertainty in the past underscores the uncertainty of our knowledge of the past.

This book explores the meanings and interactions of various uncertainties across a range of analytical scales.[61] My first concern is with scientific and climatic uncertainty as it was understood by historical actors. These uncertainties ranged from temporary gaps in climate data to "irreducible uncertainties"—enduring unknowns and mysteries.[62] On a second level, I will focus on the interconnections among scientific, environmental, and cultural uncertainties. Late nineteenth-century Euro-Americans expressed their uncertainties about climate change, environmental knowledge more broadly, and the course of capitalist development. How were these concerns related, and how can the notion of uncertainty help us better understand climatic and cultural politics in the Gilded Age United States? Last, I acknowledge the uncertainty faced by all who attempt to reconstruct and reimagine the past. Even though Gilded Age Americans wrote countless treatises and articles about climate, the historical record is marked by gaps, silences, and mysteries. I try to resist the urge to impose a tidy taxonomy on the past—the temptation to force order onto a chaotic scientific and cultural landscape, to fit complex historical actors into categories and dichotomies. In keeping with the mind-set of some climate theorists from the late nineteenth century, I hope to embrace messiness and uncertainty.

This book is by no means a comprehensive study of the cacophonous late nineteenth-century climate debate.[63] Indeed, my scope is inherently limited by its focus on Euro-American climate writings. Despite its focus within the sphere of Gilded Age Euro-American print culture, this book includes the voices of many authors who wrote letters, articles, and reports addressing possible climatic changes. Some of these writers, such as George Perkins Marsh and Gifford Pinchot, were prominent figures in their day and have since retained much of their fame, earning places in the pantheon of early conservationism. Others, such as Gustavus Hinrichs, wielded considerable influence in the Gilded Age but have since languished in near obscurity. Most, however, wrote only a single letter, article, or pamphlet about climate change before fading from the historical record. I consider all of these figures—whether powerful and famous or seemingly insignificant—to be part of the capacious category of "climate theorists." My aim is to blur the distinction between expert (or professional) and nonexpert figures, between actors in the employ of the state and nonstate or state-adjacent actors.

Hierarchies and powerful institutions certainly shaped knowledge production in the Gilded Age. The steady, if uneven, emergence of climatology as a professional and institutional "project" accorded power and prestige to some figures involved in climate debates.[64] But especially when dealing with such elusive and hard-to-prove matters as climatic changes, theorists from many walks of life often contended with professional and state actors on an even footing. Newspaper columnists and pamphlet writers sometimes used climate data and maps with the same facility shown by authors of government reports or articles in the *American Meteorological Journal.*[65] Professional researchers usually refrained from connecting their climate theories to specific ventures like land-selling schemes. Often, however, professionals waded into the messy realm of climate politics inhabited by boosters and newspaper editors. Seemingly innocuous climate reports could embroil their authors in heated discussions on expansionism and development. Some scientists worried that, since "great corporate or private interests are to be affected by its decision," the climate debate had devolved into a chaotic morass of conflicting opinions.[66] Amid this profusion of climate theories, few people could make lasting, unchallenged claims. Making matters even more confusing, climate writings showed few easily traceable patterns of change over the course of the 1870s and 1880s.[67] Even as climate change debates reflected transformations and struggles taking place across American society—industrialization, economic panics, and the growth of the bureaucratic state, to name just a few—the tenor of climate discourse remained chaotic and uncertain into the Progressive Era.

Late nineteenth-century climate writings invite comparison to our own fraught climate politics. Over the past century, climate science has evolved into a "vast machine," to use Paul Edwards's term.[68] Even with the increasing professionalization and reliability of climatology, we still sometimes struggle to transform climate science into climate politics.[69] Like Fernow and Hinrichs, contemporary people trying to grasp the consequences of climate change must deal in uncertainties and associate with unknowns. Climate models are not always as precise or accurate as we may wish, some aspects of global warming are not entirely understood, and scientists still debate how cultivation, afforestation, and deforestation can alter the regional and global climate.[70] These uncertainties need not prevent us from taking dramatic and systemic action to avert climatic catastrophes.

For decades, "merchants of doubt" have succeeded in producing uncertainty for nefarious ends.[71] In some instances it may be best to combat these tactics by erring on the side of "strategic positivism" and certainty.[72] Yet uncertainty is too powerful and protean to be ceded to these cynical actors. Perhaps we should cultivate our relationship with the unknown and carefully communicate our uncertainties. Climate debates from the past suggest how we might incorporate mysteries and unknowns into a more powerful form of environmental politics. Gilded Age writers accepted unknowns and uncertainties as intrinsic components of climate science and climate politics. Their multifaceted relationship with uncertainty holds a mirror up to our own struggle to translate scientific complexity into political action.

CHAPTER 1

A Climate Fit for Civilization

The representative from the Bureau of Indian Affairs paid a visit to Joseph Henry, a renowned scientist who inhabited an office in a "safe place in the tower of the Smithsonian." As he entered Henry's study, the visitor saw a map of the United States measuring about eight square feet. Henry noticed that the map had piqued his guest's interest and pointed out "the line which shows the region of rain; green up to that line and yellow beyond it, where they have no rain as a general thing." When the bureaucrat asked his host why he hadn't published the map, Henry replied, "because . . . Mr. Seward has wished it to remain here, lest it prevent the settlement of the West." The scientist went on to explain that, according to his data and maps, the western "prairies never had trees except along the rivers, because of the want of rain." Henry also added that any effort to plant forests to reclaim the West's climate from aridity would be futile. Unlike many of his contemporaries who blamed Native Americans for damaging the Great Plains' supposedly verdant environment, Henry believed that the "theory that the prairies were bare because the Indians had burned them off was not a true one." For Henry the map proved that the climate of the West had always been dry and would remain rainless after Native people had been confined to reservations or exterminated. The representative from the Bureau of Indian Affairs probably expressed surprise at hearing a figure as prominent as Henry question one of the tenets of expansionism—that the western landscape would reveal its latent potential only once it was settled by Euro-Americans. But Henry sounded certain and resolute in his scientific assessment of the map and of the western climate in general.[1]

The anecdote about Joseph Henry and the Smithsonian may be apocryphal. I encountered it as a secondhand account reprinted in a Kansas newspaper in 1879. Regardless of its accuracy, the Smithsonian story sheds light on the intersection between geographic or scientific debates about climate and the broader cultural, political, and economic questions facing Gilded Age Americans. If it took place at all, the exchange between Henry and his visitor probably happened sometime between 1861 and 1869, during William Seward's tenure as secretary of state. Over the following decades, figures ranging from prominent scientists to small-town Great Plains newspapermen discussed climate in light of broader cultural, social, and political questions.[2] Late nineteenth-century writers often described the stakes associated with these debates. As Kansan Frederic Hawn wrote in an 1880 newspaper article, the livelihoods of thousands of settlers in the High Plains and Intermountain West hinged on the veracity of climate change theses. Hawn alluded to a cartographic line similar to the one Henry described: "I look with anxiety upon the gloomy prospect now before thousands of the settlers in the western portion of our State. Their all is at stake. They have gone beyond what would seem to be the safe line of settlement for the present at least. It is popular to say that there is a change taking place in the climatic condition of the great plains. . . . This is either a grand truth or a most dangerous heresy."[3] Although he sounded far less certain than Henry, Hawn shared some of the famous scientist's concerns about settlers' prospects.

Writings such as Hawn's article touched on myriad topics beyond the fates of farming families and fledgling settlements. By connecting climate change to Native Americans, Mormons, ranchers, railroads, and irrigation, Gilded Age authors placed climate at the forefront of cultural contestation surrounding the project of capitalist expansion into the West. Native Americans and Mormon settlers occupied especially important places in the Gilded Age climatic imagination. For very different reasons, these two eclectic groups came to embody Euro-Americans' climate dreams and anxieties. Climate theorists understood Native Americans through an assemblage of stereotypes and tropes. Often, writers portrayed Native people and nations as a monolith, as the last major obstacle preventing Euro-American settlers from unlocking the West's climatic potential. At the same time, some climate theorists used depictions of Native Americans to criticize expansionist imperatives. Often, however, they employed the same demeaning images and vocabularies as did their more overtly racist contemporaries and opponents in the climate debate.

Mormons were never subjected to the same genocidal policies and rhetoric as Native Americans. Surprisingly, though, they occupied a similarly ambiguous and contested place in Euro-American climate discourse. Non-Mormon climate theorists viewed Mormons with both

admiration and scorn. Even as the Mormons' perceived religious and
cultural heterodoxy aroused suspicion, their seeming environmental
successes in the arid Intermountain West earned them a place as pioneers in climate improvement—at least in the eyes of some. For both
unreconstructed expansionists and those who sought to rethink the
ideology of development, Native Americans and Mormons played a
key role in attempts to imagine the climate of the West as uniquely fit
for "civilization."

Native Americans in the Euro-American Climatic Imaginary

Boomers, boosters, and some government officials predicated expansion on the notion that Euro-Americans—usually Anglo-Americans—
were the only legitimate users of land and resources. In 1889, for example, Adolphus Greely wrote a report arguing that "the confining of
Indians to reservations has removed one fruitful cause of fires during
the last ten years, so that the stunted forests are having an opportunity
of increasing the limit only by the operation of natural laws." Greely did
not feel confident enough to proclaim that confining Native Americans
to reservations would lead to a major climatic improvement. "The effect of forests as factors in the increase of rainfall," he wrote, "is still
more or less questioned." Hedging his bets, he implied that rainfall levels might "slightly" improve in the future. Despite the uncertainty of
his climate science, Greely was resolute about the theory that Native
Americans wreaked environmental havoc in the West. Greely served
as chief signal officer of the US Army Signal Corps when he published
his report on rainfall in the West. Until 1890–91, when Congress shifted
meteorological responsibilities to the newly established Weather Bureau within the Department of Agriculture, the Signal Corps led efforts
to monitor and assess the nation's weather and climate. Greely's report
indicates that genocidal climate improvement theories permeated the
highest echelons of government and science.[4]

Clearly, climate change theory played a role in legitimizing dispossession. But not all of Greely's contemporaries shared his views. Two
years before Greely published his report, the American Forestry Congress held its annual meeting in Springfield, Illinois. Many members of
Congress cast Euro-American farmers, not Native Americans, as the
primary scourge of the western landscape and atmosphere. Dan Berry
of Currin, Illinois, presented a paper claiming that the dry "air currents
that sweep over the Mississippi Valley" were in large part caused by
"the fact that the once verdure-clad western plains were yearly reduced
to plowed ground." Curry's fellow Forestry Congress member Joaquin

Miller took the argument even further. Using soaring rhetoric, he proclaimed that "10,000 iron-toothed mills," "gang-plows," and "circular saw implements" could precipitate in "two centuries what it took Babylon twenty to bring about." Instead of blaming Native Americans for ongoing cycles of "fires and floods," Miller recounted a story from his youth. He reported that the Modoc people he encountered in his childhood used controlled fires to manage and encourage forest growth: "It was my fate to spend my boyhood with the Indians. They were the only true foresters I ever met. In the spring after the leaves and grasses had served their time and season in holding back the floods and warming and nourishing the earth, then would the old squaws begin to look above for the little dry spots of headland or sunny valley. And as fast as dry spots appeared they would be burned."[5]

Miller, a famous poet and frontiersman from California, was not known for his honesty. He was such a complex character that he cannot be categorized as a critic of expansionism: penning wistful reminiscences was one thing; opposing the forces of development was another. Indeed, his nostalgic remarks fit into a broader genre of Romantic nineteenth-century depictions of "Indians" as a "vanishing race."[6] Yet Miller's paper demonstrates that some Gilded Age Americans questioned the core assumptions of the ideology of progress and development. For Miller, Native Americans were pioneers in using fires for conservation while Euro-Americans were illegitimate custodians of trees, soil, water, and air. Miller and other authors who discussed Native Americans from an environmental and climatic perspective helped bring to light undercurrents of ambivalence beneath the tide of expansionist rhetoric.

At times, however, their voices were in danger of being drowned out by Greely and others like him. Even cautious supporters of settlement often interspersed racist epithets amid their criticisms of the greedy machinations of moneyed interests. As an Englishman, Edwin A. Curley identified himself as a "disinterested observer" in the debate about Native Americans, climate, and environment. Curley toured the Great Plains in 1872–73 and wrote a lengthy study of Nebraska's natural resources. Although his book contained ads for railroads and he received payments from the Union Pacific (Railroad) Land Department, Curley acknowledged that his "era of eager enterprise and electric energy" also contained an abundance of "artful avarice." He sought to distinguish himself from unmitigated boosters and denounced pamphlets and other "crude concoctions of petty local officials who have studied no standards." In keeping with his cautious approach, Curley argued that when "tillage becomes general, and orchards and groves and forests are planted the evaporation from the soil will be more gradual," leading to reduced winds, steadier temperatures, and "probably" an "increase in rainfall" as well. Curley's stance toward Native Americans combined genocidal hatred and condescending paternalism. He argued that Euro-Americans needed to "protect Indians by putting them on reservations" and that the West "was made for men and not for red-skinned 'varmints.'" In a passage underscoring the instrumentalization of "the Indian" in Gilded Age culture, Curley invoked Native Americans while leveling criticism against robber barons: "Stealing horses is one of the principal virtues of an Indian, as stealing money, land, railroads, mines, etc., is a great virtue among some very prominent classes of whites."[7]

During the 1870s, agricultural-, railroad-, and mining-driven settlement proliferated across shortgrass prairies in the High Plains and deserts in the Intermountain West—areas that had been largely skipped over by previous waves of settlement. The boom-bust cycles of westward expansion often generated critical and cynical responses such as Curley's. Uneven settlement and economic development in these purported "virgin lands" also prompted a reconceptualizing of the relationship between Euro-Americans and Native Americans. Growing economies of extraction necessitated a stronger effort to push Native people "to the fringe," in contrast to the earlier era characterized by trading, especially in furs.[8] Climatic writings played a key role in this process of marginalization.

Climate change proponents and other expansionists employed a variety of strategies in their efforts to legitimize the Indian Wars. The most frequent involved conflating all Native Americans as a monolithic menace to forests and climate.[9] Ironically, earlier in the nineteenth century, figures such as Andrew Jackson had equated Native Americans with "unimproved" forested landscapes. In an 1830 speech on "Indian Removal," Jackson asked Congress, "What good man would prefer a country covered with forests and ranged by a few thousand savages to our extensive Republic, studded with cities, towns, and prosperous farms embellished with the improvements which art can devise or industry execute?"[10] Many Euro-Americans changed their tune in later years as expansionists' focus shifted toward the semiarid portions of the West and as dispossession policies shifted toward Native inhabitants of the Plains. By the 1870s, climatic concerns helped transform Native Americans from "savage" forest dwellers into "savage" threats to precious forests. In an 1879 testimony to Congress, for example, Nathan Meeker of Denver depicted Native Americans as careless with timber resources: "The destruction of timber is largely by fires . . . more by Indians than by white men. Indians look upon timber as of no more value than rocks, to be used if they need."[11]

Other expansionists employed subtler strategies than Meeker's. Instead of arguing that Native Americans posed a threat to nature, some authors depicted them as part of nature. "Indians" were, in the words of Nebraska booster Charles Dana Wilber, "co-workers with the natural forces that maintain and extend desert conditions."[12] During the same year that Congress published Meeker's testimony, the *Topeka Commonwealth* printed a pamphlet on climate improvement that naturalized Native Americans: "The buffalo grass, like the Indian, cannot withstand the pressure of civilization." By equating Native Americans with disappearing species, the *Commonwealth* portrayed their extermination as a natural process.[13] Viewed together, Meeker, Wilber, and the newspaper highlight the malleability of climate theory; they demonstrate how Euro-Americans deployed climate and environment to bolster both the "antiecological Indian" trope and the "vanishing Indian" narrative.

The *Commonwealth*'s naturalizing of Native Americans mirrors the work of late nineteenth-century painters who incorporated Indians into landscapes as "no more than an aestheticized 'part of nature.'"[14] Yet the writings of climate theorists indicate that the transition toward the naturalized, Romantic, and ahistorical view of Native Americans was both slow and contested. Gilded Age climate treatises sought to reconcile climatic history with competing narratives of Native American history. Sometimes this process involved drawing dubious distinctions between different Native groups. In 1880 the government surveyor and steadfast climate change proponent Ferdinand V. Hayden drew a stark line between "wild Indians" and the inhabitants of the "Colorado drainage-system." Hayden's description of the "Moqui" and Pueblo verges on the Romantic; he characterizes them as "probably the last remnants of a once great race which covered this region at one time with a great population." Hayden went on to describe how ruins in the Colorado Plateau testify to the existence of a once "peaceful, quiet, pastoral, and agricultural" society that was overwhelmed by a "rude onslaught" of "barbarous tribes from the North."[15] Even though he allowed that "few facts are known on which to base" a historical theory, Hayden's writings demonstrate how boosters and expansionists crafted a self-serving narrative under the guise of Native American history.

Hayden's comments need to be understood in the light of long debates about the origins of Native Americans and the question "Whence Came the American Indians?"[16] The surveyor's musings also fit into a well-established tradition of invoking benevolent and industrious ancient "Indians" in order to rationalize the dispossession of still-extant peoples.[17] In 1855, for example, C. B. Boynton and T. B. Mason traveled across Kansas and wrote an emotive ode to the "millions who once dwelt on these broad plains, devoted to the gentle arts of peace." Despite their admiration for the agricultural former dwellers of the Plains, Boynton and Mason did not hesitate to voice their disdain for the "wandering savage" and the "degenerate children of a wild, conquering race." These two authors believed that Euro-Americans needed to sweep aside Native people in order to restore the western landscape to the environmental and climatic splendor it had shown under the aegis of a mysterious earlier civilization.[18] The notion of re-creating a long-lost ancient civilization acquired even more power in light of widespread beliefs that the continent had once been inhabited by a "lost white race."[19]

The self-serving aspect of Hayden's and Boynton and Mason's depictions of Native Americans is apparent. Other Gilded Age authors, however, invoked historical and contemporary Native people as part of their implicit and explicit critiques of expansionism. Even while employing the same stereotypes and stories as more classic "Indian-hating" texts, reformist climate theorists used Native Americans in their attempts to imagine a more restrained Euro-American civilization in the West. A few of Hayden's contemporaries portrayed Native societies from both the past and the present as careful stewards of the climate. Some went so far as to juxtapose Native Americans' beneficial influence on climatic patterns with the deleterious impact of European societies on climate. J. H. Tice, for example, claimed that, unlike many Europeans, Aztecs and Toltecs had learned the climatic benefits of forest conservation. At the 1870 meeting of the Illinois State Horticultural Society, Tice, a resident of St. Louis, Missouri, delivered a presentation titled "Meteorological Effects of Forests." Tice deftly intertwined scientific theses with historical arguments. After explaining the role of landscape features in maintaining atmospheric "vapor in suspension," Tice delved into Aztec and Toltec history. These societies, he argued, had inhabited the Great Basin and the "plains of the Colorado" until "great climatic changes" caused by deforestation had forced their retreat to the Valley of Mexico. "They had learned a severe lesson from experience," Tice went on to explain, and had thus implemented strict "laws against the wasting of forests." He believed that Euro-Americans should mimic the "civic polity of those nations relating to forests" and asked his readers if they would be willing to atone for their environmental sins: "Will we repair the injuries the thoughtlessness and recklessness of man inflicted upon the earth and its climate? Will we go forth and ameliorate the rigors of climate on our great Western plains?"[20]

Not all of Tice's contemporaries agreed with his interpretation. In 1886, *Scientific American* published a piece discussing ruins and the once-humid climate of Arizona and New Mexico Territories; the journal claimed that neither the area's former wetness nor its current aridity had been caused by human activities. Meteorologist Mark Harrington, meanwhile, countered G. P. Marsh-inspired theories about the desiccation of the Southwest by claiming that the region had always been arid.[21] Yet Tice was far from the only late nineteenth-century thinker to invoke older societies from the Great Basin and Colorado Plateau while crafting environmental parables for contemporary Euro-Americans.[22] Unsurprisingly, most created narratives that fit their visions for the future of the West. In 1882 prominent forester Franklin B. Hough wrote that "the destruction of forests" and the concomitant climatic degradation served as a probable explanation for the disappearance of the builders of cliff dwellings and other structures in western Colorado and New Mexico.[23] The following year Patrick Hamilton, a semiofficial promoter for Arizona Territory, described an Edenic "golden age of Arizona." Hamilton's *Resources of Arizona* used images of a formerly lush and verdant landscape created by Native people to convince prospective settlers that they could restore Arizona to its earlier climatic condition. If Native Americans had succeeded in using cultivation and irrigation to ameliorate climate, the logic went, so would Euro-American farmers and irrigationists.[24]

Despite the lack of consensus among Euro-American sources about Native societies from the past, a contingent of Gilded Age writers viewed Native Americans as complex historical agents endowed with the power to improve climate. For some, such as Hough and Tice, Native history served as a warning against the excesses of expansion and heedless resource extraction. For others, such as Hamilton, Native Americans from the past served as role models in ongoing efforts to make the desert bloom. For still others, even contemporary Native Americans held valuable knowledge in the quest to better understand climate and weather. Signal Corps official W. A. Glassford gave credence to climatic theories held by Native Americans living in the areas surrounding the San Francisco Peaks in Arizona Territory. According to Glassford, "the Indians call[ed] . . . Agassiz Peak the 'mount that sits on the clouds' and Do gho slee, or weather maker" and attributed local weather patterns to the extent of snow cover on the peak. Glassford, a careful compiler of climatic and meteorological data, endorsed the notion that snow cover on mountains could exert an influence on the surrounding valleys. He explained that "when [the San Francisco Mountains] are barren of snow a change, sudden and marked, occurs," in large part because high-elevation surfaces better absorb heat and thus modify passing air currents.[25]

In his other writings, however, Glassford voiced his support for expansionism and his relief that "hostile [Native] bands have been reduced to submission."[26] Figures such as Glassford and Tice stood at the intersection of the conflicting cultural currents of the Gilded Age. They demonstrate how climate change theory encompassed both expansionist racism and a persistent strain of uncertainty lurking beneath the rhetoric of conquest.

The "Wonderful Success Achieved by the Mormons at Salt Lake"

In 1878 four cartographers working for John Wesley Powell's survey of the "Arid Region" of the United States published a map of Utah Territory "representing the extent of the irrigable and pasture lands" within the Territory (fig. 1.1). In addition to depicting wagon roads, railroads, and telegraph lines, the map used vivid colors to portray "irrigable lands," "standing timber," and "areas destitute of timber on account of fires." The four mapmakers—Charles Mahon, J. H. Renshawe, W. H. Graves, and H. Lindenkohl—did not specify who set the fires or whether the changes in Utah's forest cover would result in adverse climatic changes for the Territory. As such, the map held ambiguous implications for Native Americans, Mormons, and other groups living in the lands it depicted.

In 1882 soil and climate scientist Eugene W. Hilgard offered his own interpretation of the map. According to Hilgard, the fires that damaged Utah's stands of timber had been set by Native Americans "for the purpose of driving game" and by "carelessness on the part of whites as well." From his perspective, the map's most important message was its illustration of "the fearful depletion of the timber resources." Hilgard believed that "apart from the question of timber supply, the depletion of forests, whether by fire or by the ax, exerts a most injurious effect, in respect to both the amount and availability of the rainfall." Even though he allowed that "the former relation may be disputed by some who will not recognize any broad facts unless reducible to figures and instrumental readings," Hilgard viewed the Powell survey's map as an ominous portent for the climate and environment of Utah.[27]

In contrast to Hilgard, however, many Gilded Age expansionists believed the climate of Utah Territory had been growing wetter thanks to Euro-American, and especially Mormon, influences. For some, Mormon settlers around the Great Salt Lake served as a vanguard in climate amelioration efforts, as role models in a small-scale test case showing the potential for environmental transformation throughout the West. The question of Mormon climate improvement received so much

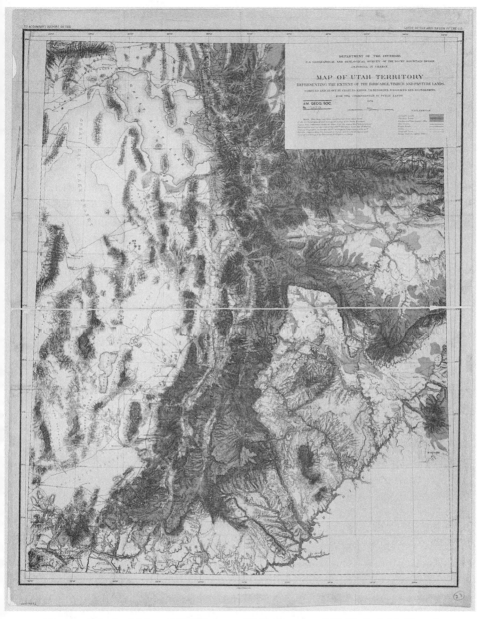

Figure 1.1. Charles Mahon, J. H. Renshawe, W. H. Graves, and H. Lindenkohl, *Map of Utah Territory, Representing the Extent of the Irrigable, Timber and Pasture Lands* (Department of the Interior, US Geographical and Geological Survey of the Rocky Mountain Region, 1878). In the original map, irrigable lands are shown as green, standing timber as blue, and burned-over forests as orange-brown. Courtesy of American Geographical Society Library.

attention that, while conducting his survey, Powell dispatched his loyal second-in-command, geologist Grove Karl Gilbert, to conduct a study on the matter. Ultimately, neither Powell nor Gilbert endorsed the notion that Mormons had been able to ameliorate Utah's climate.[28] But as historian Richard Francaviglia has observed, Powell's many maps of Utah "were meant to stimulate the development of the territory." Powell broke with many of his associates in the federal government who despised the Latter-day Saints. His "partnership with the Mormons in mapping the West stemmed less from his appreciation of their theology than from his approval of their communitarian mission to settle the land sustainably."[29] The work of Powell, Gilbert, Hilgard, and other Gilded Age writers demonstrates how climate, culture, and contestation converged in Utah. Relative to Native Americans, Mormons occupied a different place in climate theorists' visions of the West. But many authors used Mormons the same way they used Native Americans—to prove or disprove an eclectic range of theses about climate, environment, progress, and the future of the West.

Theories about Mormon-led climatic improvement centered on purported increases in the level of the Great Salt Lake. According to some, a substantial increase in the lake's level coincided with the proliferation of Mormon settlements and the growth of Mormon agriculture, proving that irrigated lands and crops attracted precipitation to the lands west of the Wasatch Range. It is unclear how this narrative about the Great Salt Lake developed or who began disseminating it, but as early as the 1860s, several non-Mormon travelers passing through Utah remarked on the lake's rising level and its climatic implications. In their 1861 narrative *A Journey to Great-Salt-Lake-City*, Jules Remy and Julius Brenchley—a Frenchman and an Englishman—set out to correct "inaccuracies" and "misrepresentations" in existing sources about the Mormons of Utah. Remy and Brenchley denounced the "mysticism" of Mormonism and labeled Joseph Smith's golden plates "instruments of . . . fraud," but they also praised the Mormons' hospitality and generosity. As for the water level, the two travelers described the Great Salt Lake as a vestige of a much larger ancient body of water that had been devastated by a "change of climate." Recently, Remy and Brenchley believed, "a change of climate in an inverse sense" had caused the lake level to rise steadily. Remy and Brenchley hoped this fact, coupled with the endorheic hydrology of the Great Basin, might "tranquilize the Mormons, who fear the lake will some day dry up."[30] Judging from the two travelers' account, settlers in Utah experienced much the same dialectic of hope and anxiety as Great Plains farmers.

Although Remy and Brenchley believed climate changes in Utah were "thoroughly proved by the increase of [lake] waters during the last few years," they did not specify how the changes had come about.[31] They did not speculate about whether increased rainfall levels could be explained by anthropogenic causes or by climatic cycles independent of human influences. Four years later another traveler, Massachusetts journalist Samuel Bowles, described potential climatic changes in Utah in even more uncertain terms. Bowles traveled across the United States with a party of journalists and politicians, among them Schuyler Colfax, Speaker of the House. Since the transcontinental railroads had not yet been completed, Bowles and his fellow travelers made part of their journey by wagon. Bowles praised the railroad as the "great creator of this empire of ours, West of the Mississippi." According to him, however, the empire created by railroads would be pastoral rather than agricultural; when the "railroad shall supersede cattle and mules," Bowles wrote, "it will feed us with beef and mutton, and give wool and leather immeasurable." Perhaps, as an easterner, he did not want to see western agriculture further supplant its eastern counterpart. It is not surprising, then, that Bowles offered only a lukewarm endorsement of Mormon-focused climate change hypotheses: "Some theorists contend that with the occupation and use of the country, the rains will multiply; and the observations of the Mormons give a faint encouragement to this idea." Bowles echoed Remy and Bretchley's ambivalence about Mormonism and Mormon-controlled Utah, but he appeared more optimistic than those two travelers. Alluding to long-standing tensions between the federal government and the leaders of the Latter-day Saints' "theodemocracy," Bowles expressed his hope that the US government could bring about a resolution through a combination of "wise guardianship" and "firm principle." After receiving a personal tour of Mormon irrigation districts from Brigham Young, Bowles sounded another hopeful note: "I find that Mormonism is not necessarily polygamy; that the one began and existed for many years without the other."[32]

By the 1870s and 1880s, early speculations like those of Bowles and Remy and Brenchley gave way to a wide-ranging and contentious debate on Mormon influences on climate. Later sources shared the tendency of early travel narratives to interweave climatic and environmental theories with cultural commentary on Mormons and westward expansion in general. The notion that early pioneers transformed the "inhospitable wasteland" of the Wasatch Front into a garden through providence and industry is well entrenched in Mormon collective memory.[33] Some non-Mormon writers adhered to this narrative and upheld the Latter-day Saints as role models in the quest to reclaim the West from aridity. In 1883, for example, forestry advocate Floyd Perry Baker cited the "wonderful success achieved by the Mormons at Salt Lake," mostly, in Baker's opinion, through irrigation.[34] William Babcock Hazen offered a more nuanced and morally ambivalent explanation for the Mormons' achievements in an 1875 piece published by the *North American Review*. "The success of the Mormons in Utah," he wrote, "has been brought about by special causes,—religious fanaticism, a mild but forcible despotism, the industrious habits brought from Northern Europe, and the spur of a lucrative market" linked to "the discovery of the precious metals in the adjoining Territories." Hazen refuted theories about anthropogenic climate change, implying that the Mormon-driven environmental transformation of Utah was a one-off that could not be repeated in other parts of the arid West: "The phenomenon of the formation and rapid growth of new, rich, and populous States will no more be seen in our present domain."[35] Whereas Hazen's arguments attest to the persistence of uncertainties about capitalist development, L. S. Burnham's 1879 letter to Congress shows that not all expansionists endorsed climate change theories. A Vermonter who moved to Bountiful, Utah, Burnham pointed out that he was a Mormon but "not a polygamist." He denied facile theses about climate improvement and attributed Utah's agricultural success to hard work: he invited others to follow his example and dig wells to access groundwater.[36]

Even climate change proponents split over the Mormon question. Hayden, so effusive in his endorsement of climate improvement caused by farmers in Kansas and Nebraska, offered little such praise for Mormon settlers in Utah Territory. In his 1871 survey report, Hayden allowed that climate "modifying influences have been put in motion in Montana, Utah, and Colorado," but his disdain for the Mormons prevented him from delving further into the topic.[37] Hayden's 1880 tome *The Great West* showed his contempt for the "arrogance of Brigham Young and his followers."[38] While Hayden's views on Mormonism stopped him from doing anything more than acknowledging Mormon-induced climate improvement, Greely—the Signal Corps chief who endorsed "confining" Native Americans—was so impressed by the "most rapid rise of the water of Salt Lake," that he felt compelled to contradict the findings of his fellow government scientist Gilbert. After praising Gilbert's "systematic and careful" study of the Great Salt Lake, Greely asserted that the lake-level increase "occurred between the years 1862 and 1870; that is to say, during the period when the amount of land being brought under cultivation and the quantity of vegetation and the number of trees was most largely increasing." Greely lamented that Utah rainfall records were "too broken . . . to show the exact relations of the rain-fall to the rising lake."[39] Despite the uncertainty of the data,

Greely found the correlation between Mormon landscape modification and climate change too compelling to ignore.

Gilbert's approach to data uncertainty differed starkly from Greely's. Circumspect almost to a fault, Gilbert wrote in his 1878 report that he considered a range of possible explanations for the seven- to eight-foot rise in the level of the Great Salt Lake, including the assertions that "the cultivated lands of Utah draw in the rain; or that the prayers of the religious community inhabiting the territory have brought water to their growing crops; or that the telegraph and the iron rails which gird the country have in some way caused electricity to induce precipitation." Gilbert's conclusions did not entirely rule out human agency—he stressed the influence of the "cutting of beaver dams" and the increased runoff caused by erosion and overgrazing. For Gilbert, though, weather patterns were too complex and mysterious for anyone to jump to conclusions about human influence. The geography of the arid West, explained Gilbert, caused natural changes in climate to be exaggerated. He also argued that "the weather of Utah is an interdependent part of the whole, and cannot be referred to its causes until the entire subject is mastered." According to Gilbert, humans could not yet grasp the "whole" because a total understanding of the forces of geography and climate remained in the domain of the "unknown."[40] Gilbert's uncertainty prevented him from endorsing optimistic plans for agricultural development in Utah and the West.

As a promotional organization, the Utah Board of Trade would seem unlikely to share Gilbert's cautious assessment of the territory's climatic improvement. The board's 1879 *Resources and Attractions of the Territory of Utah* pronounced its region the "Switzerland of America," a land with a climate "whose perpetual charms cannot be conveyed by meteorological statistics." Perhaps, from the board's perspective, Utah's climate was so superb that it did not need much improvement. The board's pamphlet allowed that "if the rainfall has increased because of the greater area of land cultivated and quantity of water diffused by irrigation as well as of the currents tapped in opening mines," the Great Salt Lake "may be expected to retain" its present high levels. At the same time, however, Utah's promoters also stated that "nothing in the meteorological register of the last four years indicates that the climate of Utah is growing moister."[41] The board's lack of enthusiasm about climate change may be rooted in its desire to entice tourists and health seekers interested in dry and salubrious climes. The Utah Board of Trade's efforts to promote the West's healthful air and climate underscores Gregg Mitman's argument about the role of health-oriented "geographies of hope" in the development of the West.[42] When "considering the sani-

tary effects of a sojourn or permanent residence in Utah," the board
argued, travelers should take "cognizance of ... its extremely dry air and
slight rainfall."[43] Ironically, western promoters and boosters sometimes
undermined climate improvement narratives while seeking to exalt the
region. Both Gilbert and the Utah Board of Trade considered existing
data too uncertain to permit firm arguments about anthropogenic climate change in Utah. Their writings demonstrate how uncertain scientific knowledge furthered competing and conflicting visions for the
future of the West.

In his report, Gilbert mentioned the notion that Mormon prayers
had brought increased rainfall to Utah. Gilbert's reference may have
been a humorous or throwaway remark, yet religion played a role in
shaping Gilded Age climatic and environmental thinking. In 1877,
Mormon "Apostle," church leader, amateur mathematician, surveyor,
and astronomer Orson Pratt delivered a sermon in Nebraska discussing increased rainfall and lake levels. J. C. McBride and J. T. Clarkson,
the editors of the periodical *Nebraska Farmer*, recounted how Pratt endorsed climate change theories but also "expressed it as his belief, that
this increased volume of water in Salt Lake was caused by the bursting
out of hidden springs upon the mountain sides—a fulfillment of some
prophecy of the prophet Joseph Smith." Although the Nebraskans cited
Pratt's description of the Great Salt Lake as proof that "the most arid
lands of Western Nebraska, Colorado, Wyoming, and Utah will one
day be as fertile as the Missouri Valley," they ridiculed most of Pratt's
sermon, especially the religiously tinged portions.[44] But many non-
Mormons shared Pratt's interest in synthesizing religious and scientific
beliefs. McBride and Clarkson's fellow Nebraskan Samuel Aughey, for
example, served both as an agricultural and climatic scientist and as a
theologian. In addition to advocating in favor of agriculture-induced
climate change, Aughey relished pointing out the lack of contradiction
between his Christian worldview and his scientific ethos.[45] Indeed, science, religion, and climate overlapped in the minds of other Gilded Age
Americans. Many who did not refer directly to religion in their writings
employed a quasi-religious moralizing tone when discussing climate.[46]
For them, climate was a moral issue in addition to a scientific one. No-
where is this sense of ethical urgency more apparent than in climate
treatises dealing with capitalism, climate, and atonement.

CHAPTER 2

Climate and Capitalism in the Great West and Beyond

In the last third of the nineteenth century, Americans struggled with an ever-changing capitalism marked by wage labor, labor unrest, and increasing concentration of capital and control in the hands of a few. Sometimes nicknamed the "Second Industrial Revolution," the "Incorporation of America," or simply the "Gilded Age," this era in the history of capitalism raised fundamental questions about the tenability of producerism and yeoman agrarianism while also spurring fears about misuse and exhaustion of natural resources.[1] Climate theories figured into Americans' attempts to bolster, reform, and attack Gilded Age capitalism. For many, belief in human-induced climate change served as a litmus test for faith in American progress. A few writers used climate change to develop a fundamental critique of capitalism, bringing to light an often overlooked tradition of radical environmental thinking. Other authors strove to allay the excesses of unchecked expansion in order to bolster the broader project of settlement and development. For these more reformist climate theorists, market forces could be harnessed into a beneficial influence on climate.

Railroad interests, one of the most powerful economic forces of the Gilded Age, provided a crucial impetus to proponents of climate improvement. Meanwhile, many authors opposed climate change theses on the grounds that they favored agricultural interests over other industries such as ranching. Economic calculus shaped Gilded Age climate debates, yet the vast range of climate theories cannot be reduced to a list of interests and factions. Support for westward expansion did not always correlate with belief in human-induced climate change. Climate change belief, scientific certainty, and faith in capital and progress did not always go hand in hand, while scientific uncertainties only sometimes dovetailed with cultural uncertainties about expansionism.

Following the thread of climate through Gilded Age culture reveals surprising fractures and alliances. Climate did not map neatly onto culture, and the lines between factions in climatic, economic, and political debates could not be neatly drawn. In addition to highlighting the most malignant strains of expansionist hubris, climate treatises can help bring to light the "moral complexity" at the core of late nineteenth-century cultural politics, especially as related to capitalism and the West.[2]

Conquest and Atonement

The notion of atonement held a strong appeal for writers who were uncertain about the social, environmental, and climatic costs of expansion but unwilling to renounce the capitalist project wholesale.[3] Atonement-minded moralizers such as Elbridge Gale monetized the costs and benefits of changes in climate. They used capitalist logic to criticize market-driven expansionism, offering an insight into the Gilded Age dialectic between profit motives and regret over the damages wrought by greed. In Gale's case, the sin needing expiation was erosion caused by careless, speculative cultivation. In 1878, Gale delivered an address to the Kansas State Horticultural Society detailing how deliberate and methodical farming techniques could undo climatic degradation caused by reckless agricultural practices. He denounced farmers who "thriftlessly exposed" soil and then explained how, if his audience followed his advice, "the whole face of the country would soon undergo a wonderful change, and we might hope that at no very distant day the climate of the whole country would be materially modified."[4]

In the moral economy of Gilded Age climate debates, atonement rhetoric like Gale's resonated with wide audiences. Reform-minded climate theorists could appeal to their readers by balancing screeds about environmental ruin with hopeful promises of renewal. Historian Craig Miner identified a similar two-sided dynamic in his study of western Kansas during the postbellum years. Miner characterized the recurring cycle of anxiety and hope among settlers as a "dialectic dance of panacea and despair."[5] This dialectic spread far beyond western Kansas settler narratives. Throughout the Gilded Age United States, climate discourse alternated between faith in progress and anxieties about failure, between the hubris of conquest and a desire to atone for the ravages created by expansion, between visions of thriving agricultural utopias and warnings about devastated landscapes and communities left in the wake of economic booms and panics.[6]

Some climate theorists articulated visions of global-scale climatic risk while still adhering to financial language and capitalist logic. In an 1876 piece discussing forest culture and climatic conditions in California, for instance, Elwood Cooper described a global climate crisis caused by unchecked expansion. "The earth is fast becoming an unfit home for its noblest inhabitants," Cooper wrote. Continued exploitation of the land would create a world of such "impoverished productiveness, of shattered surface, of climatic excess, as to threaten the depravation, barbarism, and perhaps even extinction of the species." He believed the answer to this crisis would come from realizing "that a tree or forest is an investment of capital, increasing annually in value as it grows, like money and interest." Cooper spoke from experience: he owned a plantation of fifty thousand trees near Santa Barbara, California.[7] For many large-scale horticulturalists like Cooper, capitalism needed to be harnessed and made to act as a force for ameliorating climate. Cooper believed that capitalism contained both the source of environmental problems and their solution, prefiguring the attitudes of some progressive-style conservationists from a slightly later era.

Echoing Cooper, M. C. Read and F. W. Hart crafted narratives of large-scale environmental catastrophe to give their writings a sense of urgency. According to Read, the entire Mississippi Valley was in danger of being reduced to wasteland by individualist, speculative, capitalist expansion. In an extensive 1884 government report on forestry and hydrology, he argued that "personal greed" in the form of unregulated "tillage" had already brought about land exhaustion throughout the West. If unimpeded, it would bring about "the complete *destruction* of [the West's] fertility." All was not lost, however. Read believed that "the arid lands of the West must be flanked, not attacked at the center"; he envisioned a methodical assault on the arid Plains that moved west to east from the Front Range. Through irrigation and forest plantations, his plan would transform the climate and reverse the damages caused by speculative ventures and haphazard expansion.[8] Read imagined silviculture and canal building as collaborative social projects—as correctives against individualist land use geared toward short—term profits. In practice, irrigation and forest planting proved just as susceptible to empire building and speculation as farming enterprises.[9] Yet Read's words reveal a disenchantment with individualism and profiteering—two staples of Gilded Age expansionism.

Iowan F. W. Hart shared Read's belief that the West offered Americans an opportunity to atone for their sins through climate improvement. Hart implied that droughts and floods in Ohio and Pennsylvania were punishment for indiscriminate lumbering. He worried that reduced rainfall endangered both "manufacturing interests," which depended on hydraulic power, and "sources of supply for our canals."

As in Read's and Cooper's pieces, in Hart's 1879 report, economic moti-
vations and moral atonement went together. In Hart's view, castigation
from nature in the form of droughts and floods did not require rethinking American capitalism. If anything, the prospect of climatic atonement heightened his faith in the necessity for economic expansion. He
believed that Iowa presented an opportunity for ethical and climatic
redemption: "I doubt not but the 'coming man' in Iowa will be enabled
to so manipulate the forces surrounding us, as to effectually abate, if not
remove, many of the evils now incident to humanity."[10]

If Iowa and the West in general offered the possibility of atonement,
the East showcased the misdeeds perpetrated by extractive economies.
Many authors interested in climate had recently arrived in the West.
Their articles, papers, and lectures abounded with passages connecting climatic desolation in the East to ongoing efforts to improve the
West's weather patterns. In 1888, Kansas Board of Agriculture secretary
Martin Mohler recounted a recent trip to Pennsylvania, his home state.
He described the effects of logging on the Pennsylvania landscape: "I
was much impressed by the disconsolate appearance of those ancient
hills, which in my boyhood years were covered by large and beautiful
trees, and these were the beauty and glory of the land." Mohler went on
to describe how, in the East, "the effect of the removal of forests . . . is
the tendency to destroy those atmospheric conditions which are necessary to the most satisfactory results in agriculture."[11] Over a decade
earlier, Rodney Welch had delivered a lecture in Lincoln, Nebraska, describing similar scenes of eastern desolation. Welch believed that eastern environments were beyond recovery. He called on Nebraskans to
compensate for devastated lands east of the Mississippi and then issued
an ominous warning in the form of a question: "Will the title 'Great
American Desert,' again be applied to a large portion of the territory of
this fair state?"[12] The West's status as a site of climatic renewal remained
far from certain, at least for Welch and like-minded authors.

Welch gave his lecture at the 1877 Nebraska State Fair, not a venue
often associated with somber warnings about atonement and degradation. Even booster literature from the Plains, which usually sang the
praises of its territory, contained references to climatic and environmental crises. Booster-scientists such as Samuel Aughey peppered
their writings with lamentations on the "vandal hand of man" and its
role in "destroying forests."[13] By the 1880s and 1890s, the High Plains
had experienced a series of droughts. Although expansionists advanced
climate improvement theses with nearly the same alacrity as in earlier
years, their pamphlets also began to denounce both the "impecunious
speculator" and "shiftless people"—or "agricultural nomads who infest

the country as they do every other region."[14] Many theorists acknowledged that western climates and environments had already been damaged.[15] Most critics implied that get-rich-quick schemers deserved the blame for environmental and climatic degradation. But climate-based critiques of American society's "wantonness" sometimes extended to yeoman settlers as well.[16]

In many Gilded Age climate treatises, the yeoman pioneer-farmer figured as a savior and a bringer of equable and favorable climes. Nebraska booster Charles Dana Wilber extolled the labor of the settler who "can persuade the heavens to yield their treasures of dew and rain" by "toiling with his hands."[17] Surprisingly, not even the pioneer farmer—a hallowed archetype in agrarian mythology—could bring consensus to proponents of climate improvement in the West. Divergent perceptions of yeoman farmers exposed fissures among climate-minded expansionists. Jonathan Periam, an Illinois-based horticulturist, insinuated that early settlers shared blame with hucksters and robber barons; he argued that early settlers' "first efforts were to destroy the natural grasses. This they have done without compensating therefor, while moneyed corporations were at the same time ruthlessly destroying the timber without replanting. In these respects all are alike censurable." Periam viewed the natural grasses of the prairies and plains as a climatic safeguard regulating atmospheric patterns. Without these grasses, climatic conditions would inevitably deteriorate. He proclaimed that the "old West is passing away" and urged inhabitants of the Great Plains to abandon the individualist and destructive attitude of earlier settlers.[18]

Clearly, not all expansionists viewed the West as an unblemished canvas or as a place of moral and climatic redemption. Despite the power and the appeal of atonement discourse, Gilded Age climate debates were too eclectic for any single narrative to predominate. Forestry advocate Thomas P. Roberts, for instance, argued that "human agency is possibly affecting the climate west of the one hundredth meridian" while at the same time categorically denying theories about desiccation in the East. Mohler, Read, and many others sought to transform climate change into a straightforward ethical question, but the moral landscape of the climate change debate remained as uncertain, unsettled, and unpredictable as the western climate itself.[19]

Railroads: "Voracious and Insatiable"

Atonement-minded climate moralizers often enumerated the sins of Euro-American pioneers and capitalists—deforestation, erosion, and get-rick-quick schemes chief among them. Yet they seldom denounced capitalism or westward expansion, believing that expansionism needed to be rethought and corrected, but not abolished. William Babcock Hazen undertook a more fundamental critique of expansionism. A career army officer and no stranger to controversy, Hazen was incensed by railroad interests and their attempts to portray the West as a fertile garden.[20] Hazen published an extensive study of the West in 1875. Titled *Our Barren Lands*, his piece assailed railroads and land speculators while also denying climate improvement theories. Hazen claimed he could prove "beyond all controversy" that the West's climate had been and would remain "constant." He believed that notions of climate improvement had been proliferated by inexperienced travelers who passed through the West during unusually wet seasons or years. To prove his point, Hazen cited the example of the construction of the Northern Pacific Railroad in Dakota Territory: "It will be remembered that it was during these two years [1872 and 1873] of very unusual growth that the Northern Pacific Railroad to Bismarck was built. This work was done by men who had no other experience of the seasons in Dakota, and it is no more than natural that they should honestly believe that they had seen a fair example of the seasons of the country."[21]

Hazen's strident criticism of climate change and agricultural settlement in the West caused him to become embroiled in a spat with his fellow army officer George Armstrong Custer.[22] The famous commander, Hazen believed, had erred in his efforts to promote lands in Dakota Territory by emphasizing their wetness. According to Hazen, Custer had been tricked by the same "experiences which deceived the builders of the road." Hazen added that climate patterns originated in the faraway Pacific, not in local or regional landscape features or land-use changes. Finally, he also rebuked Custer for claiming that afforestation efforts had succeeded in Nebraska.[23] Although much of Hazen's rancor originated in his disdain for Custer, he also articulated a sharp critique of the broader settlement project. The most glaring aspect of Hazen's attack on expansionism was his criticism of railroads. Having observed the workings of the Northern Pacific while stationed at Fort Buford in Dakota Territory, he denounced fraudulent railroad-sponsored efforts to sell or grant arid lands to the public, arguing that their efforts caused "misery and destitution." Hazen's attack on railroad capitalism matches historian Richard White's characterization of Gilded Age railroads as corrupt and inept. Like White, he believed that railroad builders, along with their government allies, forced ordinary people to bear the costs of their failures:

The wonder is that, in the presence of so great a failure, that [sic] there should still be found those to give further aid to this scheme.

Its originators made a most melancholy mistake in their estimate of this country. In the presence of all the facts, their scheme has been wicked beyond the power of words to express, for it successfully appealed to the poor, the lowly, the widow, and the orphan, to loan their little hard-earned savings. This fraud was enacted with impressible artfulness, with high sounding promises, supported by the name of the national government.[24]

Railroads, and especially new railroads in the West, epitomized speculative Gilded Age capitalism. They played a key role in generating and encouraging climate improvement theories that would favor settlement and expansion. In addition to hiring and supporting climate boosters, western railroads sponsored experimental "plantations" meant to demonstrate the feasibility and climatic benefits of agriculture and forest culture in the Great Plains.[25] Eminent railroad men such as Sydney Dillon, president of the Union Pacific, proclaimed that their industrial endeavors had benefited all of American society by improving the West's climate. Dillon wrote an 1891 article in the *North American Review* attacking antimonopolist critics and supporters of regulation. He praised the "law of competition" and claimed that "since the railway opened the great central and western plateaus to cultivation, the climate has become milder, the cold less destructive, and the rainfall greater."[26] Many climate theorists echoed Dillon in sycophantic tones. Patrick Hamilton, the Arizona promoter, glorified the "fiery annihilator of time and space" for heralding "throughout the land the richness of her mines, the fertility of her soil, the salubrity of her climate."[27] Similarly, Kansas climate theorist and experimental planation manager Richard Smith Elliott offered praise for the "sagacious and comprehensive designs of the Kansas [railroad] Company."[28] The railroad's influence reached beyond western booster literature: although W. B. Hazen lambasted railroads, many military and bureaucratic scientists shared Hamilton's and Elliott's views. In 1883, for example, Henry Allen Hazen, a "computer" in the office of the chief signal officer of the US Army, prefaced a report on western rainfall levels with this statement: "The following investigation is the outcome of a suggestion from Jay Gould, through Professor Baird of the Smithsonian Institution, that the building of railways in Texas and the southwest has increased the rainfall in that region." Gould was a prominent railroad speculator and robber baron.[29]

In light of the collusion between powerful railroad interests and supporters of climate change theory, it is tempting to characterize the Gilded Age climate debate as a Manichean struggle with Gould, Dillon, and their allies on one side and a few lonely dissenting voices such as W. B. Hazen's on the other. Indeed, many railroad supporters endorsed climate change theses, while skeptical climate theorists often criticized railroads. In 1888, for example, the editors of the journal *Science* attributed theories about climate improvement in the West to "agents of railroad companies," arguing that the "great railroad corporations have vast areas of land to sell in the Far West."[30] But in framing nineteenth-century climate change beliefs as ascientific myths in the service of railroad interests, historians have overlooked the fact that some climate change proponents blamed railroads for environmental and climatic damage.[31]

Few climate change partisans wrote antirailroad polemics as harsh as W. B. Hazen's. Yet several theorists who advocated for climate improvement included criticisms of railroads in their writings. Nebraskan forestry proponent J. S. Morton lamented the iron horse's "voracious and insatiable" appetite for timber products, while forester M. G. Kern blamed railroads for clear-cutting and the resulting climatic deterioration. Kern described railroads builders' "wasteful and destructive methods" and wrote that, despite their contributions to national growth, railroad interests "are also responsible for much of the hindrance to reform in the use of our forests." Neither Morton nor Kern counts as a radical critic of the railroad or of development in general; both believed in technological solutions—that, by implementing more efficient practices, railroads could become allies in the struggle to improve climate.[32] Perhaps because of the economic and political power of railroads, few climate change proponents could bring themselves to repudiate railroad interests entirely. Writings like Kern's and Morton's also testify to the simmering tensions between afforestation advocates and railroad-sponsored agriculturalists, usually allies in climate improvement and the shared cause of westward expansion. More broadly, their warnings underscore the ambivalence and uncertainty surrounding the railroad-driven arrival of industrial modernity in the West.[33]

Following Nature's Plan: Ranching and Climate

In addition to railroads, several other vast enterprises operated in the Gilded Age West. The ranching and beef industry grew over the course of the 1870s and 1880s, thanks in part to new railroads that facilitated transport and made long cattle drives unnecessary. Subject to booms and busts, the cattle industry was marked by increasing levels of speculation and financialization.[34] Ranching interests attracted scorn and criticism from antimonopolists and from proponents of agriculture- and silviculture-driven settlement. According to the ranching industry's opponents, monopolist "cattle barons" posed a threat to the continued success of yeoman family farming. In 1879, for instance, Uriah Bruner of West Point, Nebraska, pleaded with the US Congress to uphold the Timber Culture Act of 1873 and other legislation that favored agricultural settlement in the Great Plains. Bruner warned that "cattle monopolists will undoubtedly clamor for an abrogation of the land laws on the plains" and reminded legislators that "our government was established not with a view of enriching a few at the expense of many, but rather for the purpose of offering opportunities to the toiling millions to rise with the dignity of labor to a comfortable competence for himself and his family."[35] The Timber Culture Act sought to encourage timber growth and climatic amelioration in the Great Plains by granting 160 acres of land to settlers who planted a forty-acre grove and successfully maintained it for ten years. In practice, the act often served as a facade for land-speculating profiteers who wanted to gobble up land. But in the view of writers like Bruner, it represented a crucial bulwark protecting small farmers against large ranching interests.[36]

The Timber Culture Act and related theories about forest- and agriculture-induced climate improvement played a key role in the "woodsman's assaults on the domain of the cattleman."[37] Climate change proponents such as Nebraskan Charles Dana Wilber derided the "aristocratic tastes of the lords of the herds."[38] In the minds of Wilber and his allies, the veracity of climate change theses proved that ranching interests should cede the West to agriculturalists. Many climate theorists who defended agriculture and afforestation enjoyed the support of railroad corporations, belying their egalitarian and antimonopolist rhetoric. Indeed, Wilber, Bruner, and their allies never fully succeeded in painting ranchers as climate sinners and environmental villains. Ranchers and their supporters often proved just as adept as agriculturalists at combining climatic uncertainties with political economy.

Stockmen and their allies employed a variety of strategies in their effort to counter the claims of Wilber and like-minded authors. Kansan G. E. Tewksbury observed that, at least in his region, the ranching industry was egalitarian and dominated by "small holders" instead of "cattle kings."[39] Others, meanwhile, sought to defend ranchers by refuting climate change theory. Edgar Guild, another Kansas resident, wrote an 1879 piece casting doubt on the notion that "breaking a few hundred or thousand acres of sod" would bring about "any great change" in climate of the western portion of his state. Guild described landscape influences on climate as "hardly an axiom" and fused his scientific uncertainty with an uncertain economic and social outlook. According to Guild, "the western portion of Kansas can be considered as possibly agricultural to about the extent of the uncertainty of deciding where possibilities end and impossibilities begin." Any attempt to transform the region into a farming zone would be analogous to building "an uncertain empire on the ruins of another and equally important one, whose success is assured and whose limits are already sufficiently narrowed." Rather than endorsing a fundamental transformation of the High Plains environment and climate, Guild favored an industry better suited to the natural state of the semiarid shortgrass prairie. He supported ranching enterprises by acknowledging the limits of Euro-American expansionism.[40]

Capitalism and climate improvement rarely aligned in the minds of Guild and others who perceived Great Plains agriculture as "an uncertain empire." Theodore C. Henry took Guild's argument even further. A former "wheat king" who had turned against Kansas agriculture in favor of ranching and irrigation ventures in Colorado, Henry believed that, unlike farming, pastoralism was "adapted to the natural conditions of the plains." Referencing railroad-sponsored efforts to demonstrate climatic amelioration, he also took issue with experimental plantations: "I ventured to suggest that the campaigns of experiment ought to end. Attempts to battle certain natural obstacles had been unsuccessful in the past and would probably be in the future."[41] Although Henry and Guild were probably motivated by pecuniary interests, their statements evince a deep skepticism about environmental experimentation, limitless capitalist possibility, and human dominance over "nature."

Echoing Guild and Henry, other ranching advocates positioned themselves as humble followers of "nature's plan." Silas Bent delivered a speech at the "Cattle-Growers' Convention" in Saint Louis on November 18, 1884, accusing farmers and foresters of hubris in the face of nature. Perhaps the most eloquent defender of the cattle and beef industry, Bent submitted a transcript of his speech to the *American Meteorological Journal*, a publication that printed numerous treatises endorsing human-induced climate change. He implied that climate change proponents sought to defy "the immutable laws of Nature, which are not to be changed by man." Bent's speech challenged the claims of authors who believed that farmers and foresters were collaborating with nature by unlocking the West's hidden potential through climate improvement.[42]

In Bent's view, the fact that bison and other "wild browsing animals" once "infested" the Plains proved that nature intended the Great Plains to serve as a grazing region. The "untrained instinct of these wild herds," Bent claimed, "is Nature's testimony of the special fitness of these plains for pastoral purposes; and we, as intelligent people, cannot do better than to follow nature's promptings in the utilization of these lands."[43] Many expansionists equated bison with Native Americans and aridity, with some blaming bison herds for inhibiting climatic improvement by rendering the soil "nearly impervious to water" through their tramping.[44] One ranching supporter, by contrast, credited "buffalo buffs" with "probably the initiatory steps in preservation of water upon the plains." Buffalo wallows purportedly served as small water basins that exerted "a vast ameliorating influence upon the moisture of our atmosphere."[45] For Bent and like-minded authors, the bison served as crucial—if ephemeral—signs of nature's true plan for the Plains.

Bent's speech naturalized the mechanized ranching industry.[46] Nineteenth-century capitalists such as Bent envisioned and undertook the "production of nature"—to apply a theory honed by the Marxian geographer Neil Smith—supplanting the "original landscape," or "first nature," with a "second nature" designed by humans. The Chicago boosters in William Cronon's *Nature's Metropolis* "often forgot the distinctions between" these two "natures."[47] Similarly, Bent reconciled the two natures by arguing that the second evolved organically from the first. In the context of ranching and climate debates, naturalization discourse took on many guises. For many Gilded Age Euro-Americans, "Providence" stood in for "nature" and vice versa. In 1873, for instance, J. H. Beadle's *The Undeveloped West* issued an edict on God's and nature's intent: "Providence seemingly did not intend that farming should be the leading interest of the Rocky Mountain Region; its true wealth is to be found in mining and grazing."[48]

In contrast to Beadle and Bent, however, some writers staked out a middle ground in the debate about nature, climate, and cattle. Hugh Rankin Hilton described farmers and ranchers not as foes but as allies, both working in accordance with natural laws.[49] Hilton's 1888 report "Influence of Climate and Climatic Changes upon the Cattle Industry of the Plains" narrated the advent of Euro-Americans in the shortgrass prairie region. Ranchers and farmers, he believed, formed a powerful alliance working toward improving the climate. Hilton credited the "range cattle industry" with taking the first step toward climate amelioration: "The cattle-owner and his crew of hardy and courageous cowboys have placed themselves in the gap between the pioneer farmer and the hostile red man; between the retreating figures of barbarism and advancing civilization." Hilton believed that once their initial job had been accomplished, the ranchers should either move farther West

into drier areas or give way to smaller stock enterprises that could com-
plement farming. "Following a natural law, or a law of evolution," he
argued, "the encroaching civilization moistens the air, and so injures
the winter grasses, which are only properly cured by an arid climate,
that it is no longer profitable or humane to leave cattle dependent on
their own resources or ability to find a living. Hence owners of large
herds drift their cattle toward more arid climes, and their mantle de-
scends on the small herdsmen."[50]

Hilton never elucidated the exact character of the "natural law" he
discussed in his 1888 report. "Many of the present theories as to the
'causes' of the climatic changes now occurring on the plains," Hilton
predicted, "may be abandoned in the light of future investigations
and added knowledge."[51] Amid the uncertainty of Gilded Age climate
knowledge, invoking mysterious—and perhaps unknowable—natural
laws enabled both ranchers and agriculturalists to depict their contingent capitalist projects as inexorable. For Hilton, nature decreed that
cattlemen could coexist with farmers by abandoning free-range grazing
and shifting to small, fenced pastures.

George Loving shared Hilton's belief in the importance of collaboration between cattlemen and farmers. But Loving differed from Bent,
Beadle, and Hilton because he did not view "first nature" and ranching-
based "second nature" as being in a state of harmony. A Fort Worth
resident involved in the legal profession, Loving wrote an 1885 study of
"the future of the stock-growing interest in Texas." He viewed the prospects of the cattle industry as uncertain and potentially dire; Loving
identified threats to ranching interests including erosion and the depletion of grasses as well as unpredictable markets and demand. According to Loving, cattlemen should "regulate the number of cattle grazed
on any given quality of land in such a way as not to permanently damage or injure the range." His advocacy for the ranching industry did not
prevent Loving from endorsing theories about agriculture-induced climate change. Despite his belief that vast swaths of Texas would remain
"unsuited for agricultural purposes," Loving believed that Texas rainfall
levels would "continue to increase as the country develops and is converted into an agricultural country." Improved atmospheric conditions,
he argued, would benefit ranching areas adjacent to farming zones and
"greatly improve the range" and its "nutritious grasses."[52] Loving's writing demonstrates that the debate about ranching and climate cannot
simply be reduced to two warring factions—"woodsmen" and farmers
versus "stockmen." Indeed, both agriculturalists and ranching proponents relied on dystopian climate rhetoric.

Climate Dreams: Visions of Utopia in the Great Plains and Southwest

Ranching interests and agricultural partisans shared a tendency to alternate between fervent expansionism and concern about impending social, economic, and environmental crises. Usually, however, ranchers and their allies tended to be more skeptical about utopian schemes meant to transform the West. Even Loving and other ranching proponents who favored climate improvement believed that vast swaths of the West would remain impervious to climatic amelioration and thus favorable to grazing. Another group of authors, by contrast, envisioned grandiose, almost high modern, climate modification schemes. Some proposed detailed government-sponsored climate engineering strategems while others imagined vague reclamation projects. Few succeeded in implementing their plans. These unfulfilled visions for the future of the West add yet another dimension to the cacophonous climate debate. They show how climate change theory originated at the intersection of capitalist boosterism and the nascent field of regional planning. Though little more than dreams, regional transformation schemes advanced by James Humphrey, Richard Stretch, and others attest to the central role of climate beliefs in shaping the utopian horizon of Gilded Age culture.

The work of H. W. S. Cleveland demonstrates that climate change theories reached beyond the realm of get-rich-quick schemers and hucksters. Cleveland was a prominent landscape architect who designed urban parks and park systems across the United States. His 1873 book *Landscape Architecture as Applied to the Wants of the West* offered a large-scale plan for the "improvements of the land." Inspired by the forest culture arguments of G. P. Marsh and no longer content with "[laying] out some rich man's garden in the city," Cleveland sought to apply landscape architecture and landscape gardening to the vast spaces of the West. He envisioned the Great Plains as "raw material which is placed in our hands to be moulded into shape for the habitations of a nation, and such as we create, it must essentially remain for all time."[53] Cleveland wanted to avoid piecemeal, disorganized settlement of the West by instituting a cohesive plan based on forest culture. A great admirer of R. S. Elliott's railroad-sponsored agroclimatic experiments, Cleveland believed that the tree-planting component of his regional plan would lead to a permanent improvement in the climate of the Great Plains. In the last section of his book, Cleveland stated that "the labors of Mr. R. S. Elliott have thrown much light upon the subject" of forest planting and climate change.[54] Although it endorsed

Elliott's railroad-funded ventures, Cleveland's book contained a power-ful indictment of the West as it had been shaped by capitalist interests and their allies in government. His 1873 study articulated a "radical and enduring" critique of the oppressive spatial and social conditions cre-ated by capitalist development.[55] Perhaps Cleveland shared some of his contemporaries' desire for atonement. His treatise reveals how even ambitious regional transformation plans sometimes originated from a place of uncertainty—uncertainty about whether the costs and conse-quences of expansion outweighed its benefits.

Like Cleveland, Judge James Humphrey, chairman of the Board of Railroad Commissioners of Kansas, proposed a grand plan for trans-forming the landscape and climate of the High Plains. His 1887 pro-posal contained the same element of ambivalence present in the land-scape architect's work. Humphrey began by acknowledging the limits of human agency, arguing that the primary cause of climate patterns—great ocean currents—remained far beyond the reach of human influ-ences. Instead of stressing the importance of forest culture, Humphrey envisioned a series of artificial lakes. The judge believed that one-acre ponds, constructed in sufficient numbers, would secure "more equa-ble precipitation of moisture." Though he encouraged homesteaders to do their part and construct climate-improving ponds, Humphrey also urged the federal government to construct "a system of reservoirs upon the upper waters of the Platte," a river system encompassing parts of Nebraska, eastern Colorado, and eastern Wyoming. "The creation of . . . conditions favorable to the prosperity of millions of human be-ings," he wrote, "would be ample justification for such an outlay of pub-lic money."[56]

Proposals similar to Humphrey's appeared throughout western print culture in the 1870s and 1880s. F. M. Clarke's plan for reservoir construction in the Colorado Rockies followed the same format as Humphrey's. Clarke argued that "it is simply folly to indulge any hope that the 'rain-belt' will ever visit the 'arid region' [the Plains] so long as the Rockies rear their tall crests." He then asserted that creating moun-tain reservoirs in Colorado would prompt the "westward march of the much-prayed-for rain-belt."[57] Clarke's statements may seem contradic-tory or paradoxical, but in late nineteenth-century climate writings, limits and limitless possibilities often went hand in hand. The Gilded Age tension between humility and hubris encompassed both climate discourse and protomodernist regional planning. At the same time, the scale and optimism of Clarke's, Cleveland's, and Humphrey's visions foreshadowed twentieth-century regional improvement schemes such as the 1930s shelterbelt program.[58]

As a potential site for regional climate improvement schemes, only the desert Southwest rivaled the Great Plains. At various points in the decades following the Civil War, boosters and government officials recommended flooding swaths of desert in order to bring about a change in climatic conditions.[59] In 1874, civil engineers J. E. James and Richard H. Stretch conducted a government feasibility study "on the practicability of turning the waters of the Gulf of California into the Colorado Deserts and the Death Valley."[60] Their report detailed the potential drawbacks of such a plan, but prominent figures and expansionist authors seized on the irrigation scheme's utopian possibilities. John C. Frémont, Republican politician and former explorer, served as governor of the Territory of Arizona from 1878 to 1881 and proposed constructing a canal that would create a vast inland sea in Southern California and western Arizona. Frémont's plan received an endorsement from L. P. Brockett, whose 1881 book *Our Western Empire* blamed Native Americans for having "diminished the rainfall" of the West. Alluding to a mysterious civilization that had once inhabited the region, Brockett couched the ambitious irrigation scheme in the language of restoration. Frémont's plan, he wrote, would "restore" both "the great inland sea which formerly existed in Southern California" and the wet climate that had characterized the ancient Southwest. Brockett claimed that "evaporation from that sea would ensure a moister atmosphere and a greater rainfall to western Arizona and, in connection with other measures" such as tree planting, "would render the Territory the garden-spot of all the West."[61]

As in Utah and the Great Plains, however, conflicting interests undermined ambitious climate improvement plans in the lower Colorado watershed. Many critiques of Brockett's and Frémont's proposal originated from fellow capitalists and expansionists who believed the artificial inland sea would harm other ventures and assets. Stretch, for example, raised the question whether "it would be evidently wiser policy to retain the land than to destroy it by submersion."[62] Whereas Stretch viewed Arizona lands as possibly too valuable to submerge, some California opponents of the plan perceived the region's arid climate not as a liability but as an advantage. Echoing the Utah Board of Trade, the *Los Angeles Times* warned that increased humidity and the resulting "moist, 'sticky' heat" might prevent people from moving to Southern California.[63] While the land- and humidity-based critiques carried clear implications, perhaps the most puzzling statement about the inland sea proposal came from J. E. James, the civil engineer who wrote the 1874 feasibility study along with Stretch. James neither endorsed nor denounced the irrigation plan. After writing that "it is reasonable to suppose" that climate benefits would result from the artificial sea, James described contemporary understandings of climate as too uncertain to allow for any guarantees. He encouraged a "careful investigation to determine the correctness of the theories advanced" as rationales for creating the inland sea.[64] It is unclear whether James sought to encourage studies like his own or whether he intended to cast doubt on the entire artificial sea enterprise. Yet the southwestern climate improvement schemes demonstrate how, in addition to supporting climate theory, expansionist interests and scientific uncertainty sometimes combined to cloud transformative visions of environmental improvement.

Like the Great Plains, the borderlands simultaneously represented the promise of limitless growth and a reminder of the limits of development. Climate debates offer a telling glimpse into the unresolved struggle over the meaning of the Southwest that took place over the course of the 1870s and 1880s. Earlier in the nineteenth century, the borderlands' "arid climate, rough topography, lack of navigable waterways, and unfamiliar Hispanic Catholic society" had prompted some Anglo-Americans to view expansion in the Southwest with caution and suspicion.[65] Some of these sentiments persisted into the Gilded Age. In his study of the US–Mexico borderlands, Samuel Truett has argued that, in the Gilded Age borderlands, "the best-laid plans of states, entrepreneurs, and corporations repeatedly ran aground." Brockett, Frémont, and other supporters of the inland sea plan fall into the category of "borderland dreamers" Truett identified. These "dreamers" often betrayed their own angst about stagnation, retrenchment, and failure even as they hatched grandiose schemes.[66] Often, though, a sense of environmental and climatic exceptionalism emboldened Anglo-American expansionists who looked to the Southwest.

Ardent expansionists often categorized inhabitants of the "Spanish" settlements alongside Native Americans from the Plains as illegitimate users of the environment. From these authors' perspective, the fact that people had been practicing agriculture and irrigation in parts of Mexico, California, Arizona, and New Mexico for centuries with no discernible climatic improvement did little to disprove theories of human-induced climate change. New Mexico booster Elias Brevoort believed that Mexican and Hispano settlers had been unable to tap the latent potential of the landscape. His 1874 treatise *New Mexico: Her Natural Resources and Attractions* lamented that "the population of New Mexico hitherto has not, unfortunately, been of the progressive kind. The Spanish and Mexican Race . . . has caused the country to progress scarcely a move in the march of material improvement and wealth." In Brevoort's racially hierarchical imagination, only Anglo-Americans were capable of bringing social, economic, and climatic progress to New Mexico Territory and adjacent arid regions.[67] He claimed that Anglo-American settlers would be more methodical and careful in implementing forest culture, horticulture, and irrigation: "It is gathered from well tried experiments that, *when more attention has been given in this section to the planting of fruit and forest trees, the climate will be materially changed* [Brevoort's emphasis]." In addition to his faith in forest culture, Brevoort shared another trait with some of his contemporaries from Kansas and Nebraska: he was committed to building railroads and sought to encourage the construction of a transcontinental railroad along the thirty-fifth parallel in New Mexico.[68]

Stephen Dorsey offered a more nuanced take than Brevoort's on railroad building in New Mexico. Dorsey presented his views in an 1887 piece in the *North American Review*. A staunch Republican who owned land in New Mexico Territory and had served as a US senator for Arkansas, Dorsey advocated a reformist approach to capitalist development. He argued that his colleagues in the federal government had gone too far in granting vast swaths of land to railroad corporations. He also diverged from Brevoort in that he viewed the Anglo-American settlement project as being on par with Spanish, Mexican, and Mexican-American efforts. After criticizing the excessive scale of some railroad land grants, Dorsey explained, "On the whole . . . the railroad land grants were for the best interests of the whole country, but no more so than were the grants given by the Spanish and Mexican governments of large tracts in New Mexico, Arizona, and California, to induce colonization in some cases, and to reward eminent public services in others." Considering his validation of earlier settlement efforts in the Southwest, it is not surprising that Dorsey differed starkly from Brevoort on the climate issue. Dorsey posed a biting question to proponents of silviculture- and agriculture-induced climate change: "New Mexico, parts of Arizona, and the Republic of Mexico have been under cultivation for three hundred and fifty years. . . . [W]hat climatic changes have occurred? Are not irrigation canals required now as in centuries gone by?"[69]

Dorsey fit the profile of the most maligned opponents of agriculture and climate change theory: being involved with mining and ranching claims and interests, he supported those industries in New Mexico; he voiced his opposition to the continued use of "ordinary farming" in the "arid region" under the 160-acre allotment plan of the Homestead Act; and, most damningly, he echoed John Wesley Powell's argument that the region's dryness necessitated rethinking the settlement of the West. Economic and political exigencies may have prompted Dorsey's opposition to climate change theory, but his views reveal that not all expansionists espoused Brevoort's brand of race-based climate belief and not all Gilded Age authors shared Brevoort's view of the Southwest as a potential Eden for Anglo-American capitalists.[70]

Like Dorsey, T. C. Henry, the former Kansas "wheat king," considered the possibility of climate change in light of older Spanish and Mexican settlement projects. He believed the borderlands contained a lesson in limits for Anglo-American settlers, who, he implied, were no better than their Hispano counterparts. In a pair of lectures delivered in Kansas in 1882, Henry recounted his experiences from recent trips to Mexico and New Mexico. He relayed anecdotes about his meetings with two prominent men—Trinidad Romero, a former delegate to Congress from New Mexico, and Governor Luis Terrazas of Chihuahua.[71] While with Romero, Henry consulted "records preserved in his family, reaching back to the Spanish settlement of the Territory—more than two hundred years ago," which showed "indisputably that [New Mexico's] forestry, rainfall, and general climate features are unchanged." He found similar evidence of climatic continuity in Chihuahua, where Governor Terrazas proved that "large areas have been irrigated, and agriculture maintained for centuries" without discernably "modifying climatic results." Henry's consultation with Romero and Terrazas affirmed his belief that settlers in the West should adapt their economies to fit environmental conditions rather than seeking to transform climates and landscapes to fit their needs. Addressing his Kansas audience, he explained that "the present physical phenomena of the plains and prairies of Kansas will continue practically unchanged, and every successfully organized industry must be conformed to them."[72]

Climate, Science, and Culture

For Dorsey and Henry, opposition to climate change theory dovetailed with opposition to unmitigated capitalist development. Dorsey's and Henry's work points toward a persistent uncertainty about the tenability of unchecked capitalism. As for scientific uncertainty, however, neither author discussed the topic: Dorsey summarily dismissed climate change belief as "idle" talk, while Henry sounded resolute and certain in his conclusion about climate stasis.[73] Judging only from these two sources, it would be tempting to view scientific, economic, and cultural uncertainties as unrelated. Indeed, some strains of climate change theory originated among shrill and confident boosters who merely used scientific uncertainty to serve their ends. At the same time, however, other Gilded Age writers invoked uncertain scientific knowledge while voicing their skepticism about heedless expansion. Late nineteenth-century climate theorists fused climatic politics and cultural politics to the point of inseparability, precluding any straightforward conclusions about the relation between cultural uncertainty and scientific uncertainty. While climate discourse helps reveal the dialectics between conquest and atonement and between dreams of utopia and visions of decline, it does not elucidate the precise nature of the connection between uncertainty, science, and Gilded Age expansionism.

The chaotic and confused character of late 1800s climate politics poses a challenge for any effort to use precedents from the 1870s and 1880s as models for contemporary climate communications. Yet the contested and muddled aspect of late nineteenth-century climate writings might resonate with twenty-first-century climate thinkers. Today we still struggle to reconcile political economy with potential climate futures.[74] Some contemporary authors view climatic progress and capitalism as irreconcilable. Others echo atonement-minded Gilded Age authors who sought to harness market forces and transform them into a beneficial influence on climate.[75]

In light of these diverging opinions, the malleability of historical climate discourse and its intricate, shifting relationship with culture might offer a useful precedent. The contentious political environment of the Gilded Age, coupled with the still-inchoate character of climatology, forced writers to meld economic prescriptions with uncertain climate theory. Whether in the pages of government reports, in newly formed scientific journals, or in popular newspapers, climate theorists brought the contingencies of science together with the contingencies of capitalism. By invoking mysterious "laws of nature," they infused their political and environmental visions with a sense of urgency. Instead of emphasizing scientific truths unsullied by politics, it may be more advantageous to emulate historical climate theorists by focusing on the intersection of science, culture, and political economy. The notion of "nature's plan" may no longer carry the same currency or resonance as it did in the late 1800s. But Gilded Age climate theory underscores the importance of incorporating scientific, cultural, and economic visions to craft an accessible form of environmental politics.

CHAPTER 3

In the Middle Border

Gustavus Hinrichs and His Network of Volunteer Observers

During the Gilded Age, in the states and territories of the interior United States, Euro-Americans interested in establishing weather and climate observation networks faced numerous obstacles. Securing funding ranked high among them. In 1882, for example, Francis Nipher helped introduce a bill before the Missouri State Legislature seeking several thousand dollars of funding for the Missouri Weather Service. "Consideration of this bill," Nipher later wrote, "furnished an occasion for mirth to some members of that body; but failed to occasion any interest." The Saint Louis resident found the rejection "so depressing" that he made "no further attempt . . . in that direction." Yet Nipher did succeed in expanding the Missouri Weather Service, thanks in large part to a growing network of volunteer observers—weather aficionados who recorded rainfall and temperature data to be collected in reports such as an 1892 study titled "Missouri Rainfall." According to Nipher, the "patience and self-denial" of volunteer observers allowed his weather service to overcome the indifference of state legislators.[1]

State weather services similar to Nipher's proliferated across the "middle border"—a vaguely defined region spanning parts of the Midwest and Great Plains—during the 1870s and 1880s.[2] These local or regional-scale networks reflect some of broad trends in the history of climatology and meteorology. In the United States and beyond, increasingly systematic, methodical, professionalized, and quantified disciplines supplanted weather folklore and climate studies carried out by polymaths. Indeed, members of regional weather bureaus perceived their work as contributing to a "system" of "increasing complexity and abstraction" to be used by scientists in support of "theoretical, polemical, and practical objectives."[3] Local networks also supported the development of "agricultural capitalism" throughout the West.[4] And they participated in the "messy and incomplete" transition toward more standardized data-collection frameworks.[5] Yet the idiosyncrasies of these local midwestern and Great Plains networks merit closer examination. The inner workings of organizations such as Nipher's highlight eddies amid the seemingly inexorable currents of nineteenth-century science: the shift from settler ecologies and folklore to data-based epistemologies, and the drive to separate "pure science" from the murky realms of boosterism and cultural politics.[6]

Climate theorists from the "middle border" region sought to derive usable climatic knowledge from sprawling sets of data collected by observers. But nineteenth-century data collection efforts rarely gave rise to stable and certain scientific paradigms. Attempts to standardize meteorological and climatic data ushered in a "period of reevaluation and uncertainty" over the direction of climatology as a discipline.[7] Often, accumulating vast troves of numbers such as those Nipher compiled created more questions than answers.[8] Data collection efforts failed to resolve the politically charged issue of whether Euro-American settlement could modify climatic conditions through agriculture, afforestation, deforestation, and other means. Despite their inability to definitively answer climate change questions, Nipher and his contemporaries created a syncretic form of science, a science that fused naturalistic beliefs with quantitative methods and probabilistic paradigms with experiential and experimental modes of apprehending the natural world.

This chapter examines the dialectic between institutional science and everyday science.[9] In the late nineteenth-century United States, boundaries between types of knowledge remained porous and contested. Quotidian settler perceptions, folklore, and natural inquiry did not simply disappear in the face of bureaucratized, professional science.[10] Middle border boosters and newspapermen sometimes deployed, debated, and questioned expansive data sets. "Men of science," meanwhile, used anecdote and folklore as starting points in deriving hypotheses to explain climatic and meteorological changes. In some cases, experts sought legitimacy from their use of experiential and practical knowledge.[11] I treat newspapermen, boosters, volunteer observers, and highly trained academics as exponents of a similar form of syncretic meteorology and climatology. These figures, I argue, did not so much create a discrete branch of knowledge as operate within a network of mutually influencing epistemologies. The ability to navigate among different audiences, different forms of data, and different ways of knowing made some Gilded Age climate theorists uniquely adept at practicing the uncertain art of climate politics.

Gustavus Hinrichs, Mercurial Polymath

Gustavus Hinrichs, the founder and head of the Iowa State Weather
Service, stood at the intersection of the divergent intellectual currents
shaping Great Plains meteorological and climatic science. At times he
upheld the work of "practical" men over that of institutional scientists,
railed against the dogma of positivism, and advanced boosterish, promotional climatology. But at other times he policed the borders of science, issuing vehement denunciations of anyone straying beyond the
bounds of "real science." Hinrichs also made seemingly conflicting
statements about human-induced climate change and about the role of
statistical uncertainty in climate science.[12] Aside from occasional credits for coining the term "derecho" (a large, straight-line windstorm), he
has received little attention from historians of climatology and meteorology.[13] A mercurial and paradoxical figure, Hinrichs offers a glimpse
into contentious Gilded Age debates over who qualified as legitimate
creators of climatic and meteorological knowledge. His work merits
closer examination because, despite its idiosyncrasies, it is representative of "middle border" knowledge making. By combining experiential
and statistical evidence, Hinrichs managed to cope with the myriad
problems facing Gilded Age climate theorists: growing but incomplete
and unreliable data sets, multiplicities of often contradictory hypotheses, and growing, but occasionally hostile and alienating, bureaucracies.
    Hinrichs was born in 1836 in Lunden, a city then located in Den-
mark, and immigrated to the United States in 1861 because of political
turmoil surrounding Prussian unification. First settling in Davenport,
Iowa, he then moved to Iowa City in 1863 to work as a professor at
the State University of Iowa, where he taught foreign languages and
physical sciences (fig. 3.1). A polymath, Hinrichs conducted research
in mineralogy, meteorology, medicine, geology, physics, and chemistry, earning some renown for his work on periodic laws.[14] In 1875 Hinrichs established the Iowa State Weather Service and later took pride
in having organized what he termed "the *first* State Weather Service
of America."[15] His weather service operated continuously until 1889,
when Hinrichs moved to St. Louis and his network was supplanted by
the rival Iowa Weather and Crop Service, an organization affiliated with
the US Weather Bureau.[16]
    A mixture of personal, utilitarian, and theoretical goals motivated
Hinrichs's efforts to develop the Iowa State Weather Service. He envisioned his weather service and data-gathering network as a project
undertaken by the people of Iowa, for the people of Iowa. By collecting
temperature, wind, and rainfall statistics, he argued, the weather ser-

Figure 3.1. Gustavus Hinrichs. Courtesy of the Department of Special Collections and Archives, The University of Iowa Libraries.

vice would "secure a faithful record of the conditions on which Iowa's prosperity depends and will continue to depend."[17] In his *Biennial Reports*, Hinrichs stressed the importance of producing meteorological information of immediate use to agriculturalists and others working in his adoptive state. While long-term climatic observations would inform the ongoing process of farming-based settlement, short-term meteorological studies would help Iowans understand, cope with, and potentially predict destructive storms and tornadoes. Hinrichs's 1877 report offered a rousing defense of the Iowa State Weather Service. Data collected by his network, he claimed, had "conclusively demonstrated" that timber areas exerted a strong influence on the "amount, frequency, and intensity, as well as the distribution of fertilizing thunder-storms." Hinrichs viewed weather patterns as dynamic and as susceptible to human agency—mostly through afforestation and deforestation—so he believed Weather Service "results" should "form the basis of rational legislation . . . having for its object the increase of healthfulness and fertility of entire regions of our State."[18] In other instances, however,

Hinrichs characterized the data supporting theories of human climatic influences as "apparently contradictory." But he presented these doubts and contradictions as proof that his Weather Service should continue to carry out its work "by extended observation and reduction."[19]

Hinrichs depicted his work as both a utilitarian endeavor and a noble, esoteric pursuit driven by curiosity about the unknown. He fused "everyday science" with "pure science," seeking immediate material rewards while also uncovering mysteries and raising new questions about the sources of environmental and climatic changes.[20] *Rainfall Laws Deduced from Twenty Years of Observation*, published in 1893, represents the culmination of Hinrichs's work in Iowa and his efforts to meld citizen science and practical knowledge with complex statistical methods. In this work, Hinrichs outlined a series of logarithmic equations for determining the relative agricultural utility of rainfall events. His experiences working in his "large garden" near the "bluffs of the Iowa River" piqued his interest in juxtaposing the success of agricultural endeavors with rainfall statistics provided by volunteer observers. Noting the inadequacy of simple precipitation totals and the influence of evaporation and runoff, Hinrichs wrote that the "thrashing [*sic*] machine seemed to be entirely independent of my rain gauge." As an alternative to simple rainfall numbers, Hinrichs devised a series of "laws" and parameters that would transform meteorological statistics into dynamic tools in the service of agriculture. Categories such as "total utilizable rains" and "total useless or damaging rains" would support efforts to assess crop prospects. It is not clear whether Hinrichs's efforts succeeded in rendering multitudes of new weather statistics more useful and legible to farmers. Still, *Rainfall Laws* highlights Hinrichs's belief that climatic laws and meteorological statistics would only beget uncertainty unless paired with material, quotidian realities.[21]

Although the US Weather Bureau published Hinrichs's statistical tome, the Iowan had a contentious relationship with the Washington-based scientific establishment. In 1891 the civilian US Department of Agriculture took over the US Army Signal Service's weather reporting network. The national network had been growing throughout the 1870s and 1880s, sometimes collaborating with local and state-based organizations such as Hinrichs's. In Iowa, however, scientific and personal conflicts prevented smooth cooperation. George E. Curtis, a prominent federal bureaucrat and climate theorist, published a scathing review of *Rainfall Laws*, calling various portions of the book "obscure" and confusing. Curtis also took issue with Hinrichs's purportedly "unnecessary" discussion of probability.[22]

Hinrichs cast similar aspersions on the work of the "national weather bureau" and defended his choice to establish an independent state weather service in Iowa. As if their "indifferent, if not hostile" attitude toward the Iowa State Weather Service were not bad enough, Hinrichs wrote, the US Army Signal Corps, the Weather Bureau, and the Smithsonian Institution also carried out shoddy science. In an 1887 report, he admonished the national weather bureaucracy for emphasizing "the production of so-called indications and probabilities" to "the detriment of real climatological study." The Iowan added that he hoped "a broader, a more scientific spirit" would "in time prevail in the management of the national weather service."[23] Though Curtis and Hinrichs both viewed uncertainty as an unavoidable component of scientific theory and practice, they accused each other of overreliance on "probabilities." The issue of probability stirred controversy across realms of Gilded Age society from actuarial science to daily weather forecasting.[24] At times climate theorists gained power and prestige by showing their mastery of contingent forms of knowledge. As shown by the Hinrichs-Curtis spat, however, dealing in uncertainties and "probabilities" also exposed climate theorists to criticism.

The conflict between Hinrichs and national institutions may have arisen not just from intellectual disputes, but also from his sometimes pugnacious personality. During his time at the State University of Iowa, Hinrichs clashed with administrators and fellow professors. In a decade-long effort to have the polymath dismissed, other faculty members at the university presented the Board of Regents with lists of grievances against Hinrichs. They claimed that he belittled his colleagues "in his classrooms, on the street, at home, and abroad," even "to the extent of using profane language." Other disputes centered on Hinrichs's purportedly excessive salary and his supposed appropriation of university equipment, including an "electric lantern" and a heliostat.[25] Although in 1886 his colleagues succeeded in having Hinrichs dismissed from the Collegiate Faculty, students and several local newspapers came to his defense. The *Iowa City Post*, for example, decried the "most desperate and dastardly assaults upon the good name of Dr. Hinrichs." The climate theorist lamented his dismissal, saying he was "neither invited nor allowed to defend" himself.[26] In the decades following his firing, many Iowans and others sought to rehabilitate Hinrichs. A 1923 obituary indicates that his allies and admirers did succeed to some extent in reclaiming the polymath's reputation. Its author, Charles Keyes, enumerated Hinrichs's scientific achievements in climatology, chemistry, and crystallography while adding that the Iowan had been "coldly received" in his home state but had garnered "loud applaudits everywhere throughout intellectual Europe."[27]

Hinrichs's personal squabbles notwithstanding, disagreement over

the proper scale for conducting climate research drove his clash with national scientific institutions. Late nineteenth-century efforts to centralize
and "scale up" observation networks often raised new questions about
the relation between lay knowledge, local beliefs, and "universal" scientific knowledge.[28] As sociologist Phaedra Daipha has argued, data collection projects "intensified jurisdictional wars . . . over the merit of local
weather versus local atmospheric systems, observation versus speculation, reportage of unusual weather versus global atmospheric systems."[29]

In the Great Plains, some scale-based "jurisdictional wars" centered
on extreme weather events. Tornado forecasting, and even the use of
the panic-inducing word "tornado," provoked sometimes bitter contention in the late nineteenth century.[30] Hinrichs had a special interest in
"destructive great storms," especially tornadoes. In an 1889 article, he
published a map showing all recorded tornado tracks in Iowa from 1875
to 1888, asking government scientists to "stop the manufacture of dire
tornadoes"—in other words, to cease issuing exaggerated reports about
tornado dangers in Iowa.[31] Only local experience, he argued, would allow for a proper tracking and warning system. After describing a system
of "weather flags" meant to communicate barometric changes to the
community, Hinrichs stated, "It is of supreme importance that our people should learn to help themselves, and not vainly rely upon a distant
power which even at best cannot reach them until too late. Weather
telegrams are of greatest possible value, but only as aids to properly
organized local work."[32] New technologies, he insisted, would be effective only if implemented on a local scale and in conjunction with
local knowledge. As to whether Iowa "is big enough for a weather service," Hinrichs answered, "indeed it is," because it was far bigger than
"England, Portugal, Switzerland," each of which had separate, state-
supported weather services.[33]

Though he invoked other nation-states to legitimize his work, Hinrichs and his weather service cannot be cast as simple vehicles for
state-driven modernization and centralization. National weather prediction systems and all-explaining theories of storm formation and climatic changes did not inspire Hinrichs as much as the task of keeping a
"faithful exposition of the actual conditions of the weather in Iowa, so
that our Weather Reports will continue to be of value long after views
and theories shall have passed away." Hinrichs sought to "simplify and
systematize" institutions and flows of information. In 1870, he founded
the *American Scientific Monthly* and stated that the journal would act
as "an exponent" of "modern science," "the spirit which is fashioning
this age."[34] Yet he espoused a capacious brand of modern science, one
that included polymaths and resisted the hardening of discrete disci-

plines.[35] Perhaps Hinrichs's expansive interests made it hard for him to find a niche in the growing national scientific bureaucracy and meteorological network. The friction between the Iowa State Weather Service and the Washington, DC–based scientific establishment also underscores the constant contestation and renegotiation between "centers" and "peripheries" in Gilded Age scientific practice.[36] Data did not simply flow from the interior to nodes of knowledge production on the Eastern Seaboard and then, reconstituted as "science," trickle back to the peripheries.[37]

While resisting encroachment from the national metropolis, Hinrichs created his own center of calculation in Iowa City.[38] His house in Iowa City served as the "Central Station" of the data collection network.[39] Hinrichs's home featured a three-story tower, its top two floors dedicated to the Weather Service, and a rooftop balcony complete with weather vane, thermometer, hygrometer, and rain gauge (fig. 3.2).

Though Hinrichs conducted his own observations, his most arduous task was managing his network of volunteer observers. By 1877, Hinrichs wrote, he could count on "eighty-seven volunteer observers representing as many *Stations*." I was unable to find archival evidence about these citizen scientists, their views on science, climate, weather, and politics, or about the nature of their relationship with the weather service's founder. But the fact that Hinrichs termed each observer a "station" indicates that he accorded some measure of respect and gratitude to the volunteers upon whom he depended. The Weather Service's second annual report (1877) included a map showing the location of each volunteer observation station (fig. 3.3).

In his reports and articles, Hinrichs offered a glimpse into the challenge of creating an imagined community of science over such an expansive territory.[40] He described Iowa as a settler society marked by transience: "In our comparatively new State people change residence more frequently than in older states." Volunteer observers dropped out because of death, disease, or "neglect." Adding insult to injury, he complained, the national Weather Bureau "attempted to estrange our volunteers" during the 1880s. Despite these difficulties, volunteers provided Hinrichs with a flood of data. He lamented the dearth of "clerical help" and spent long hours mailing, copying, and "office printing." In 1878 alone, he claimed, "44,502 copies were made from 166 stencils."[41] Creating tables, maps, and reports from data provided by volunteers was another herculean task. "It should be remembered," Hinrichs wrote, "that in this work there is no cessation; every day brings its load of facts and data which have to be properly classified, recorded and disposed of."[42] Hinrichs created a system of forms intended to facilitate corre-

Figure 3.2. Central Station of the Iowa State Weather Service, Iowa City. From Hinrichs, *Second Biennial Report of the Central Station of the Iowa Weather Service* (1882).

spondence with observers and ease the translation of data from the field into charts and eventually maps.[43] The cartographic process, however, remained tedious. For example, an 1883 series of maps correlating timber areas with rainfall averages contained a staggering 26,082 rainfall measurements.[44] The work proved so onerous that Hinrichs wondered if it would be "imprudent" for him to continue his "personal sacrifice" and "expenditure of labor and money."[45]

Despite his rhetoric about advancing "modern science," Hinrichs sometimes agonized that his work might be "thoroughly useless."[46] With the popularizing of science across the United States, new journals and climatic theories proliferated, creating a cacophony of voices.[47] The contributions of the Iowa State Weather Service risked being lost in this chaotic scientific cauldron. Hinrichs reassured himself and his volunteers with the hope that "every true observation made by any of our observers at any station in Iowa will . . . constitute an additional link in the chain which binds the past to the future."[48] Since Hinrichs found hope in the notion of a growing web of climate knowledge, he may have been reassured to find that in 1893 Corydon P. Cronk of the Maryland State Weather Service cited his contributions: "In the state of Iowa it has been conclusively proven, by the records of the State Weather Service, that the annual rainfall is more evenly distributed throughout the year in the more heavily wooded portions of the state." Bolstered by data from Iowa, Cronk made strident claims about forests' influence on climate patterns. He even speculated that afforestation and reforestation might offer "protection from the tornado" by preventing the "overheating of the earth's surface" and thus diminishing the "energy of these storms and . . . lessen[ing] the frequency of tornadoes."[49] Cronk likely derived his inferences from cartographic series created by Hinrichs (fig. 3.4), perhaps the map series that required over 26,000 observations. Yet his claims maintained none of the uncertainties and qualifications that appeared in Hinrichs's work.

Though the Iowan did pronounce that he had "conclusively demonstrated" forest influences on climate, he often followed his statements with calls for further research.[50] Cronk's use of Hinrichs's research offers

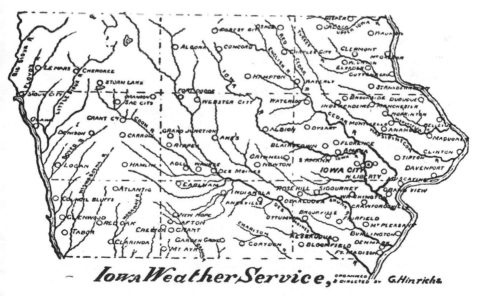

Figure 3.3. Stations of the Iowa State Weather Service. From Hinrichs, *Second Annual Report of the Iowa State Weather Service* (1877).

Figure 3.4. Cronk may have been referring to these maps (or to similar map series made by Hinrichs) in his 1893 article. Writing about the maps described above, Hinrichs claimed that the "distribution of the shading expressing the amount of rainfall (in inches) shows a close relation to the distribution of the shading marking the percentage of the surface covered with timber." Thus, he argued, the maps "furnish abundant material support" for the theory that society could influence climatic patterns through afforestation and deforestation (624). Image and quotations from Hinrichs, *Second Annual Report of the Iowa State Weather Service* (1877).

a glimpse into the circuitous networks of knowledge circulation in the Gilded Age United States. The diffusion of information and its interpretations was not one-way: climate theory sometimes flowed from west to east along with data. And the reuse and reframing of information at each center of calculation sometimes added layer upon layer of uncertainty.

Syncretic Science

Historians have sometimes characterized theories of local and regional forest-climatic influences as "mythological conceptions," the last vestiges of naturalistic, folkloric, and pseudoscientific paradigms.[51] Yet Gilded Age climatological and meteorological writings from the "middle border" offer more evidence for continuity and syncretism than for a straightforward transition from naturalistic beliefs to data-derived scientific knowledge. As David Livingstone and Charles Withers have observed, nineteenth-century scientific thinkers inhabited multiple spaces and "operated different moral and epistemic economies."[52] Cronk, Hinrichs, and others employed both quantitative and anecdotal, observational evidence. After invoking Hinrichs's statistical studies, for example, Cronk remarked, "The traveler who now crosses the continent through the states of Iowa, Kansas, or Nebraska will see the strong belts of forest trees which the laws of the states have compelled the owners of land to plant. The results have been marked. The rainfall is more evenly distributed."[53] Cronk and Hinrichs employed different lexicons to engage with different audiences in seeking financial, institutional, and moral support. Despite their syncretic approach—and at times because of it—Hinrichs and his contemporaries took part in "boundary work," the strategic practice of attempting to exclude other authorities from the scientific realm.[54] Some proved adept at a peculiar juggling act: policing the boundaries of science while also working to "translate" between different ways of knowing.[55]

In addition to polymaths and institution builders such as Hinrichs, other Euro-Americans engaged in a vigorous debate over the proper parameters of meteorological science. S. L. Dosher, an observer for the national Weather Bureau who was based in Manistee, Michigan, found fault with the persistence of naturalistic impressions in climatological studies. In 1893 Dosher wrote a letter to the *American Meteorological Journal*, a periodical that published contributions from prominent government scientists as well as Hinrichs and similar figures. Vague impressions drawn from hazy memories, Dosher insisted, could only give rise to fallacious climatic theories. "Whenever there occurs a period of extreme heat, a long wet spell or dry spell or even a period of exceedingly fine weather," he wrote, "people will always claim that no such weather ever occurred before." Dosher alleged that such misconceptions engendered the "general opinion prevailing that the climate of our country is changing, especially with reference to the winters, which, it is often claimed, are growing milder." Dosher found no evidence of climate change, anthropogenic or otherwise, in the records of multiple Weather Bureau stations.[56]

Some "middle border" climate theorists shared Dosher's belief that only data collection could offer definitive solutions to scientific quandaries. In 1878, for example, Kansan Isaac Noyes wrote that he believed efforts to collect "daily facts" would "enlighten mankind with the mysteries that preside over the natural phenomena that govern the weather." Quantitative triumphalism, however, did not always foster consensus on the contentious question of human-induced climatic changes. Unlike Dosher, Noyes allowed for the possibility of climatic changes and supported the notion that society could influence weather patterns.[57] The decade and a half that elapsed between the publication of Noyes's and Dosher's pieces cannot entirely explain this difference of opinion, since Hinrichs and many others employed data-based approaches to support theories of anthropogenic climate change well into the 1890s.[58]

Like Dosher, some Great Plains climate theorists sought to purge climate discourse of what one Iowa horticulturalist termed "moonshine notions."[59] At the same time, climate-related newspaper stories such as the *Topeka Daily Tribune*'s 1879 "Bogus Science against Experience and Common Sense" show that some Great Plains Euro-Americans valued "practical knowledge" over "science."[60] Much of this hostility toward high science and its exponents originated in resentment of eastern elites, a widespread sentiment among middle-border climate theorists.[61] But a broad cross section of Gilded Age scientific thinkers attempted to incorporate folkloric beliefs within quantitative methodologies such as those employed by scientists in eastern metropolises. Some Great Plains polymaths, horticulturists, newspapermen, and university "men of science" seemed to take their cue from John Trowbridge.

An easterner and professor of physics at Harvard, Trowbridge wrote an 1872 piece in *Popular Science Monthly* arguing that "great fires" have "with some probability of truth . . . an influence upon the production of rain." He derived his hypotheses from folkloric notions about rainstorms following fires and admonished colleagues who dismissed folk beliefs out of hand: "The attitude of scientific men in regard to so-called popular fallacies and superstitions is not, in general, a praiseworthy one. A belief needs only to be widespread among the people at large to be denounced." Trowbridge conducted a series of electrical experiments in his laboratory in an effort to simulate the effect of fires on atmospheric conditions. Though unable to rule out uncertainties arising from his methods, he found that the experiments affirmed naturalistic impressions that large-scale fires triggered rainstorms.[62] As a proponent of laboratory experiments, Trowbridge participated in an ongoing debate about the relationship of natural philosophy, pure science, applied science, and engineering. On the one hand, contestation over the boundaries of these disciplines gave rise to purity discourses— efforts to expunge purportedly illegitimate epistemologies. On the other hand, it created openings for people such as Trowbridge to draw from folklore and popular impressions.[63]

Trowbridge used popular beliefs and anecdotes only as a starting point, a means of devising a hypothesis to be tested in a laboratory. Some middle-border climate writers, by contrast, considered evidence drawn from experience and observation alongside evidence obtained by measurement. In the work of J. L. Budd, a member of the Iowa Board of Forestry, experiential evidence filled gaps and voids in the statistical record. Budd presented a paper titled "Possible Modification of Our Prairie Climate" at the 1887 meeting of the American Forestry Congress in Illinois. His presentation sought explanations for recent crop failures in Iowa. "Ordinary meteorological tables," he argued, "are not sufficiently detailed to throw light on the influence of the weather on agricultural and horticultural crops." Making no apparent reference to the efforts of his fellow Iowan Hinrichs, Budd described Iowa's statistical record as too brief to reveal anything more than "*probable* causes for *known* effects." He relied on personal experience for proof of these "*probable* causes." Ironically, Budd's experiences and observations prompted him to identify plowing as the cause of Iowa's agricultural and climatic troubles. Since vast swaths of the state's land had been "turned with the plow," winds "from all westerly points now literally pass over a dry heated soil in a dry period, which drinks up with hungry avidity the moisture of the air." According to the Iowan, "methodical forestry planting" could act as a "complete or partial remedy" for the "climatic troubles" plowing created. The crux of Budd's argument—the causal factor at the core of the purported "modification" of climate—originated from naturalistic impression and observation. In the absence of empirical evidence, he relied on his own experience on the land to prove that plowed soil could draw moisture from passing air masses. For Budd, experience and anecdote infused some certainty into the probabilistic frameworks of statistical climatology.[64]

Great Plains climate theorists invoked popular impression and the weight of experience for varying purposes, sometimes to endorse theories of human influence on climate and sometimes to cast doubt on such notions. In 1878, for example, William Tompkins of the *Larned (KS) Press* cited anecdotal evidence of dew formation to show climatic continuity. The newspaper claimed dew formed with as much frequency in 1873, when Native Americans still "roamed over the land," as in succeeding years, after Euro-Americans had plowed thousands of acres of soil near Larned.[65] As Tompkins's piece shows, the eclectic range of Gilded Age climate discourse cannot be distilled into a simple dichotomy: folkloric proponents of climatic improvement or desiccation versus quantitative modernizers who refuted climate change theses.

Perhaps no figure better reflects the syncretic character of Great Plains climate science than Francis Huntington Snow (fig. 3.5). Snow's views on climate mirrored Budd's theories more than those of his fellow Kansan Tompkins. A polymath with interests ranging from entomology to botany to meteorology, Snow published a series of articles on climate and climate change from the 1870s to the 1890s. Though his career in some ways paralleled Hinrichs's, Snow proved more adept at climbing institutional hierarchies than his contemporary from Iowa. He began teaching mathematics and natural sciences at the University of Kansas in 1866 and rose to be chancellor in 1890. Like Hinrichs, Snow earned greatest renown outside the fields of meteorology and climatology, garnering recognition for his discovery of a fungus useful in combating chinch bugs, a scourge on agriculture.[66] Snow's wide-ranging interests informed his approach to climatic questions: he used different methods and sought to reconcile experiential evidence with statistical records. In 1873 Snow trumpeted

Figure 3.5. Francis Huntington Snow. Courtesy of
University of Kansas Archives (F. H. Snow Clippings
File).

the "self-registering instruments" and "automatic [apparatuses]" used by
the University of Kansas's meteorological station. Although he supplemented data from the university observation station with numbers from
Smithsonian Institution observers, Snow believed that it would take fifty
years of weather records to determine the accuracy of climate change
hypotheses. In the absence of such statistics, Snow made recourse to
popular sentiments. He viewed naturalistic impressions as a stopgap but
implied they carried inherent weight and authenticity, especially when
attributed to the "oldest residents of Kansas."[67]

Snow believed that Euro-American settlement had increased rainfall
amounts in Kansas through a variety of means, ranging from plowing
to preventing prairie fires to replacing "short buffalo grass" with "longer
and heavier grasses."[68] Lacking statistical proof for increases in annual
rainfall, Snow offered experiential evidence that human agency had
rendered his state's climate more equable. In an 1871 letter to his fellow Kansan C. C. Hutchinson, Snow argued that "it certainly would
be legitimate to cite the evidence of many of our 'old settlers' to the
fact that the rain fall is more evenly distributed now than ten years ago,
coming at shorter intervals and more gently, and that single storms,

or showers, extend more hours than formerly." "This belief," Snow added, "I have often heard from our most intelligent citizens."[69] For a primarily agricultural society like 1870s Kansas, rainfall distribution mattered nearly as much as annual rainfall totals. As Hinrichs observed in *Rainfall Laws*, brief and violent rainfall could do more harm than good. The allure of reliable climates could act as a strong enticement for prospective settlers, and Kansans like Snow had an interest in attracting more emigrants and development to their state. Indeed, it is telling that Hutchinson published Snow's assessments in *Resources of Kansas*, a document meant to attract agricultural settlers. He may have been more willing to gesture toward settlers' perceptions in a general audience publication than in a more formal scientific publication. Yet Snow cannot be dismissed as a booster-scientist or a mere huckster. In an 1885 piece, Snow tempered optimistic expectations for human-induced climatic improvement. He described society's influences on climate as "local oscillations" and voiced his skepticism about the notion that Euro-American settlement would entirely transform semiarid regions such as western Kansas.[70]

In the same 1885 article, Snow identified a frustrating aspect of data-driven climatology. He lamented that the US Army Signal Corps had used incomplete records to question possible climatic changes in the Great Plains. Snow seemed to grasp some of the problems at the core of Gilded Age science: quantification created an unquenchable thirst for ever more data, and sets of numbers could be deployed to prove any number of theories. Faced with these obstacles, theorists like Snow created a holistic form of climate science that included folklore, anecdote, and experience. The fusion of quantitative and qualitative methods allowed Snow and some of his contemporaries to cope with the uncertainty inherent in meteorology and climatology.[71]

By the 1890s, however, Snow began to change his approach. At the 1895 annual meeting of the Kansas State Board of Agriculture he presented a paper titled "Periodicity of Kansas Rainfall and Possibilities of Storage of Excess Rainfall." After twenty-seven years of consistent observation and measurement at his station in Lawrence, Kansas, Snow claimed to have found a regular seven-year repeating pattern of dry and wet periods. The notion of seven-year cyclicality resonated with Great Plains climate theorists during the 1890s, perhaps because of its biblical parallels. The severe droughts of the late 1880s and early 1890s, which followed a series of wet years, may also have helped give rise to cyclical climate theories. Despite the popularity of the concept of climate cycles, Snow argued that "the common people failed to recognize the periodicity of rainfall" before remarking that "Eastern meteorologists have called attention to a similar periodicity."[72] His shift away from folkloric and experiential evidence attests to the changing nature of Great Plains vernacular science. Impressionistic evidence continued to shape climate discourse long after the turn of the century, yet it took on an ever more peripheral role, underscoring the crystallization of scientific disciplines and the standardization and quantification of meteorology and climatology in the early twentieth century.

Snow's 1895 presentation also bore the hallmarks of turn-of-the-century progressivism: concerns with utilitarian efficiency and careful resource management. Building reservoirs, he wrote, would allow Kansas to store rainfall from wet years "in such a way as to be of service in the following months or seasons when the precipitation is below average." Snow speculated that these storage reservoirs might also "increase the humidity of the atmosphere" and "reduce to an injurious minimum" damaging hot winds.[73] Hinrichs's and Cronk's reports certainly prefigured Snow's progressive turn toward efficiency. But the Kansan's later work reflects a shift away from sweeping efforts to induce and catalog climatic changes and toward a potentially systematized management of climatic variability. Snow never implemented his plan for reservoir construction. Despite its biblical resonance, his theory of seven-year cycles did not gain much traction beyond the Great Plains. Still, his shift away from experiential vernacular science reflects the increasing, if incomplete, marginalization of folkloric climate discourse around the turn of the century.

"The Creature of Climate"

Snow's adoption of progressive utilitarianism shows that "middle border" scientific syncretism involved more than just the fusion of quantitative and qualitative methods. Other intellectual currents shaped Gilded Age climate science. Medical geography and enviroclimatic determinism, for example, found their way into Snow's work.[74] In an 1876 essay titled "Climate and Brains," published by the Kansas Academy of Science, M. V. B. Knox invoked Snow's climatic expertise: "It has been suggested by Prof. F. H. Snow, that the general dryness of the atmosphere in Kansas may prove favorable to brain-workers." Knox also explained how countries in areas with propitious climates, especially those in northern Europe, had surpassed other areas in terms of cultural productions.[75] According to Knox's and Snow's logic, only Euro-Americans benefited from Kansas's salubrious climate, or else Native American inhabitants of the state would have eclipsed them in intelligence. In some instances, as in Snow's and Hinrichs's work, medical geography appeared alongside theories of human-induced climate improvement. Hinrichs, a cautious proponent of forest-induced climate improvement, cited the influence of changing climates on "the state of health of the body and mind" as a rationale for supporting "special institutions for . . . accurate observation," such as his own Iowa State Weather Service.[76] Yet Hinrichs's stance vis-à-vis climatic influences on society differed from those of Knox and Snow. His theories of climatic dynamism and his support of complex climate improvement theses prevented him from endorsing simplistic and deterministic climate theories.

Though not all Gilded Age vernacular scientists endorsed them, deterministic climate theories such as Knox's proliferated throughout Gilded Age culture. In *The Mississippi Valley*, a triumphal and expansionist book intended for popular audiences, J. W. Foster echoed Knox by writing that "however much he boasts of his dominion over matter, [man] is the creature of climate."[77] By depicting Euro-Americans as the sole beneficiaries of climatic influences, writings including Knox's and Foster's served to legitimize capitalist expansionism as well as the dispossession and genocide of Native Americans. The political and cultural implications of climatic determinism underscore the imbrication of science and politics in the Gilded Age. As David Singerman has argued, in the late nineteenth-century United States, increasing numbers of people realized that "scientific knowledge, far from being the inevitable ally of accountability and good governance, could just as easily be deployed to obfuscate and confuse, and thereby to wrest control of social and economic power."[78]

In a sense, the syncretic and eclectic character of Gilded Age climate science may have encouraged the strategic obfuscation Singerman describes. In another sense, perhaps, it may have flattened social hierarchies and allowed more people to participate in producing and contesting scientific knowledge. Throughout this chapter I have tried to emphasize continuity, the persistence of experiential, anecdotal science, and the messiness and false dawns that marked local, participatory data collection projects like Hinrichs's. But the story of meteorology and climatology in the Gilded Age middle border remains incomplete. Even embattled and sometimes reviled personages such as Hinrichs wielded far more influence than volunteer observers. The voices of Hinrichs and his rivals still dominate those of farmers, agriculturalists, and others who contributed as much to the project of vernacular science as did bureaucrats and polymaths. At the same time, however, the writings of figures like Hinrichs and Snow offer a fleeting glimpse into an intricate scientific universe that has largely gone unrecorded.

CHAPTER 4

Fluid Geographies
Mapping Climate Change

J. T. Allan and F. P. Baker refused to draw lines in the sand, or for that matter, in the lush prairie soil. In an 1883 government report on forestry, Baker argued that "in all the country between the Mississippi and the Rocky Mountains . . . no man can yet say where the line is located beyond which forestry is unprofitable."[1] During the same year, Allan published a railroad pamphlet about Nebraska that extended Baker's argument to include other crops as well. "With the increased yearly rainfall moving westward," he claimed, "it is not possible to fix the limit of agricultural production."[2] Baker "led a peripatetic life," working at various points as a farmer, blacksmith, attorney, canal agent, and newspaper editor before examining European forestry in his capacity as "an additional U.S. Commissioner" to the 1878 Paris Universal Exhibition.[3] Allan, another forestry enthusiast, served as president of the Nebraska State Horticultural Society.[4]

Both men supported railroad- and agriculture-driven expansion into the Great Plains. They believed that, through the spread of farming and silviculture, Euro-Americans had transformed the climate of the Plains and rendered it more conducive to settlement. By developing a dynamic, uncertain, and fluid geographical imaginary, Allan and Baker sought to counter both long-standing theories about the existence of a "Great American Desert" and contemporary cartographic visions like Charles Sprague Sargent's map of North America, which drew fixed lines separating green, forested regions from barren and treeless areas seemingly hostile to settlement (fig. 4.1).

Theodore C. Henry, a former wheat grower and a determined opponent of climate amelioration, disagreed with Allan and Baker. Although he supported the broader project of national expansion and development, he disputed the notion that there had been "any material increase

Figure 4.1. "Position of the Forest, Prairie, and Treeless Regions of North America."
Note the way Sargent labeled much of the area in contention—including western Kansas, western Nebraska, and eastern Colorado—as treeless. C. S. Sargent, *Report on the Forests of North America*, Tenth Census (Washington, DC: Department of the Interior, 1884). Courtesy of American Geographical Society Library.

in the average annual rainfall." Henry had abandoned Great Plains agriculture because he considered the region dry and thus better suited
to "graziers and shepherds."[5] Surprisingly, however, Henry did not
adopt the fixed geographical imaginary exemplified by Sargent's map.
He believed that even though the Plains were too arid for large-scale
agriculture, the region's constantly changing climate and environment
could not be captured by unmoving lines. In response to accusations
that he was creating an impassable border on the map of Kansas, Henry
wrote an article in 1882 claiming that he "did not draw an absolute isothermal line. I cannot nor can anyone else. The climatic differences are
too imperceptibly defined for that."[6] Like his seeming opponents Allan
and Baker, Henry held that cartography was incapable of portraying the
uncertain climates and landscapes of the Plains.

Despite Henry's, Allan's, and Baker's belief in the limits of cartography, maps played a crucial role in nineteenth-century debates about
climate change, settlement, and westward expansion. When drafting
their maps, Gilded Age climate theorists made use of data from networks such as Hinrichs's Iowa State Weather Service. They also employed temperature and rainfall data collected by the US Army Signal Corps and land grant universities.[7] Some mapmakers, including
Charles Dana Wilber and Hinrichs himself, used these data to create
new and dynamic cartographic methods that reflected their belief in
constantly shifting human influences on climate. Their maps testified to
society's power to transform and improve both landscapes and weather
patterns. Other climate theorists, meanwhile, continued making maps
that espoused a geography of limits. Maps such as Sargent's advanced
a different cultural, environmental, and economic vision for the West.
Sargent, a prominent arboriculture professor at Harvard University,
was, according to historian Donald Pisani, "one of the staunchest critics of the idea that forests changed the climate." His maps emphasized
the limits of Euro-American expansion as well as society's inability to
alter environments to fit its needs.[8]

But as Allen's, Baker's, and Henry's writings show, the Gilded Age
struggle over geographic imaginaries was neither straightforward nor
easily framed in terms of fixed categories. Sometimes expansionists
and climate change proponents drew lines on the map of the West, and
sometimes more skeptical climate theorists employed dynamic and fluid
geographies. As critical cartographers and other scholars have pointed
out, cartography is a "slippery" technology.[9] Few cartographers could
control the meaning of their maps. In some instances, climate change
proponents reinterpreted maps made by more skeptical cartographers
in order to bolster their arguments. Late nineteenth-century climatic

and environmental theorists constantly reshaped and recast the meaning of maps, contributing to the culture of uncertainty surrounding
questions of human agency, climate change, and expansionism.

The boosters, surveyors, and cartographers involved in the climate
debate employed a dizzying array of novel cartographic techniques.
Nineteenth-century cartographers had far more data at their disposal
than their predecessors. To portray this information, they devised a
new form of data-heavy mapping that historian Susan Schulten calls
"thematic mapping." These thematic, statistical maps included choropleth maps, isopleth maps, and point symbol maps.[10] Cartographers also
experimented with maps that defied easy categorizing, such as hybrid
maps with both isolines and point symbols. In the words of historian
Katharine Anderson, "widespread experimentation with the visual
presentation of scientific information" characterized middle to late
nineteenth-century cartography. Anderson also argues that the proliferation of new cartographic conventions and techniques led to "divergence" and fracture in the field of meteorology.[11] Although Anderson
focuses on short-term meteorological maps, her argument also applies
to longer-term climate maps such as those used to depict change or
continuity in the climate of the American West. The piecemeal adoption of new cartographic techniques underscores a paradox inherent
in the development of modern cartography and climate science: in addition to generating facts and scientific certainties, efforts to map and
monitor environments also created fragmentation and uncertainty.

My aim is not to deny that maps served as powerful tools of territorial and social control. As Schulten has pointed out, in the nineteenth-
century United States, thematic maps "captured complexity and
concretized the abstract." They played a crucial role in the "quest for
government control" that transformed national spaces and environments by rendering them legible and measurable. Historians have
demonstrated how capitalist development and westward expansion
"endowed cartography with a responsibility to demystify the West."[12]
Still, in some instances the power of maps rested in their ability to mys-
tify the West.

Scholars have shown how "counter-maps" can sow doubts and spatial uncertainties, thereby disrupting statist efforts to assert control
over local landscapes and people.[13] In the context of Gilded Age debates about climate, environment, and expansion, even maps made
by powerful interests and experts sometimes functioned by imparting
a sense of uncertainty. Some cartographers blurred boundaries and
lines rather than fixing them. Others admitted the tenuous nature of
their geographic claims and the ambiguity of the data used to derive

their maps. Denis Wood has described the map's "disguise"—the cartographer's attempt to elide any uncertainty and present a veneer of
objectivity that masks the contestation and complex social processes
that go into making a map.[14] Perhaps the "disguise" of certainty is not
the only source of cartography's power. Climate maps of the American
West reveal that cartography sometimes functioned through a dialectic
between certainty and uncertainty.[15] Some Gilded Age cartographers
conveyed and created spatial information while also acknowledging the
limits of scientific knowledge and the persistence of unknowns.

Delineating Deserts

Many cartographers, surveyors, and scientists envisioned the debate
about human influences on climate as a struggle over Americans' mental maps. Few settlers participated in making actual maps, but as Martin
Brückner has observed, the nineteenth century saw a rapid growth in
the geographic literacy of nonelite people.[16] Especially in the early to
middle nineteenth century, the "Great American Desert" dominated
many Euro-Americans' mental maps of the Great Plains and Intermountain West.[17] Sara Robinson was a New Englander who settled
in Kansas. Her 1856 book on her adoptive state offers a glimpse into
the making of the midcentury geographical imagination. Robinson re-
called learning about the Great American Desert from school atlases as
a young student:

> Most vividly of all comes before me the bright-colored map, in
> green, red and yellow, upon which I daily learned my lessons, as to
> our whereabouts, and that of mankind generally, upon the face of the
> old earth. Very many were my speculations as to the appearance of
> one part of the country, laid down upon the map as the Great Ameri-
> can Desert. There was mystery to me in its semi-circular lines in fine
> letters, "Great American Desert, inhabited only by savages and wild
> beasts," and much childish curiosity was excited thereby.[18]

In the Gilded Age, as more and more railroads and Euro-American
settlers moved into the shortgrass prairie of western Kansas, western
Nebraska, and eastern Colorado, the supposed existence of the Great
American Desert inspired not just curiosity but acrimonious debate.

Proponents of human-induced climate change believed that the
"Desert" needed to be eradicated from the nation's imagined geography. They resisted the idea that anyone could draw a line across the
prairie separating wet and fertile regions from desert areas unfit for cul-

tivation. Melville Landon—who wrote under the nom de plume Eli Perkins—drew much of their ire during the early 1880s. In 1879, Perkins had written a widely reprinted column arguing that "an awful trap is being set for settlers" by railroad interests and other boosters. Perkins wrote that the arid regions of the West would remain dry "until the Almighty changes the course of the winds, takes down the mountain-peaks, and stops the clouds from raining all their water out in the East before they get to the desert." Perkins's column prompted vociferous attacks against the notion that anyone could draw a "drouth line" across the West.[19] The *Topeka Commonwealth* took issue with Perkins's idea that agriculture would be unfeasible beyond the one hundredth meridian: "This 'One-Hundredth Meridian' idea is as fallacious as was isothermal line fancy on the [slavery] question." The *Commonwealth* was alluding to older theories that slavery could not expand beyond certain north-south isolines because of unfavorable climatic and environmental conditions.[20] By equating Perkins's thesis with ideas about slavery, the newspaper implied that he was a purveyor of nefarious and outdated geographical fallacies. In 1881 L. P. Brockett's tome *Our Western Empire* heaped more scorn on Perkins, decrying his "deplorable ignorance" before arguing that cultivation and the reclamation of arid lands could transform the climate of the West and expunge the Great American Desert from the nation's map.[21]

For many ardent supporters of development, any line on a map of the Great Plains and Intermountain West was an abomination. Yet some proponents of westward expansion and climate change tried to stake out a middle ground in the debate about deserts and isolines. Cyrus Thomas, a prominent government surveyor who believed that marginal areas could be reclaimed by climate change, allowed that certain regions would remain arid and off-limits to agriculture. Thomas wanted to find the precise "agricultural, climatological, and physical . . . dividing line" that marked which areas were closed to settlement.[22] Kansan H. R. Hilton shared Thomas's belief in the continued existence of a "dividing line." For Hilton, however, the line was not permanent. In an address before the Scientific Club of Topeka, Hilton asserted that "there is a much stronger argument in favor of a movable than of a fixed line."[23] Similarly, Kansas forestry advocate Martin Allen envisioned a constantly shifting "imaginary line somewhere away out . . . beyond the center of population in the State, where large herds of cattle and flocks of sheep are more numerous, and plowing more scant, and, therefore, prairie fires more frequent and wide-spread." Allen believed this line was "varying and uncertain."[24] His writing underscores the influence of dynamic cartography on nineteenth-century environmental thought.

Like Thomas, Hilton, and many others, Allen read and conceived of landscapes in cartographic terms, by imagining isolines.

Despite the efforts of writers like Brockett, the "Great American Desert" did not disappear from all the nation's imagined and printed maps. But Allen, Thomas, and many others helped transform the desert from a static geographical entity into an elusive place whose extent and location could be shifted and negotiated. Many mid-nineteenth-century mapmakers labeled vast swaths of the West the Great American Desert. Following in the footsteps of explorer Stephen Harriman Long, who helped coin the term in the 1820s, they used large lettering to mark much of the High Plains with this inauspicious place-name. By the dawn of the Gilded Age, however, the desert had migrated. While some cartographers working in the 1870s and 1880s abandoned the term entirely, others labeled smaller regions of the Intermountain West as a new Great American Desert. Charles Roeser's 1879 map of Utah, for example, named a portion of Western Utah the Great American Desert (fig. 4.2). Roeser's map, compiled for the General Land Office of the Department of the Interior, testifies to the fluidity of Gilded Age geographical imaginaries.

Fluid Geographies

Some proponents of climate change and westward expansion did more than contest place-names on existing maps. They experimented with thematic maps in order to demonstrate that human influences had modified weather patterns and shifted isohyetal lines in the Great Plains and beyond. Before the advent of animated maps, cartographers struggled to map change over time.[25] Indeed, many nineteenth-century mapmakers aimed to convey a sense of fixity and stasis rather than change. Charles Dana Wilber and Gustavus Hinrichs sought to inject a sense of dynamism into cartography by juxtaposing multiple maps. Their map series portrayed climate patterns as too uncertain and fluid to be captured by any single map.

Wilber was one of the best-known proponents of anthropogenic climate change in the Great Plains. A steadfast supporter of agricultural development, Wilber collaborated with his fellow Nebraska booster-scientist Samuel Aughey over the course of the 1870s and 1880s.[26] The two argued that plowing up semiarid lands had transformed weather patterns by allowing soils to evaporate more moisture into the air. Although several scholars have analyzed Wilber's writings on climate and agriculture, his maps have largely avoided academic scrutiny.[27] In 1881 Wilber criticized cartographers and surveyors who "stereotyped" the

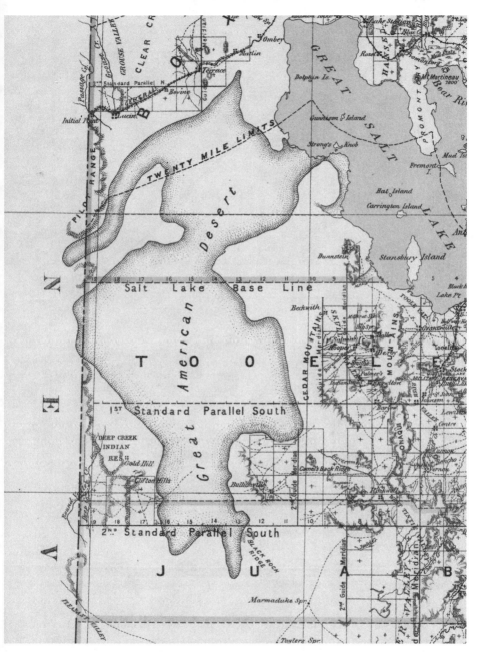

Figure 4.2. Detail from "Territory of Utah," by C. Roeser, principal draughtsman, Department of the Interior, General Land Office, 1879. Courtesy of David Rumsey Historical Map Collection.

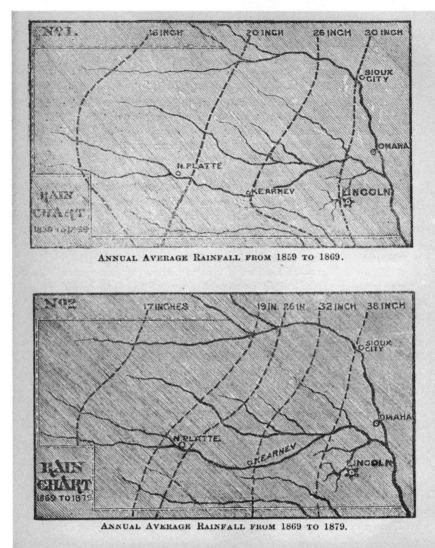

ANNUAL AVERAGE RAINFALL FROM 1859 TO 1869.

ANNUAL AVERAGE RAINFALL FROM 1869 TO 1879.

Figure 4.3. C. D. Wilber and S. Aughey, "Rain Chart," from *Great Valleys and Plains of Nebraska and the Northwest* (1881), 77.

Great Plains environment using static maps.[28] As an alternative, Wilber published two maps that, when compared, revealed the westward movement of isohyets across Nebraska (fig. 4.3). The first map depicts average rainfall levels from 1859 to 1869, while the second shows rainfall data from the following decade. The maps are somewhat difficult to read at first because they use different data ranges for every rainfall line (16, 20, 26, and 30 inches for the first map and 17, 19, 26, 32, and

35 inches for the second map). Yet the two maps still convey the transience and uncertainty of the Great Plains environment. The isohyets are marked by faint dotted lines, suggesting the ephemeral nature of climate patterns and, perhaps, the uncertainty of the data used to create the maps. Wilber's map demonstrates that Nebraska's rainfall levels had supposedly increased, but in itself it does not explain the exact causal mechanism behind this transformation.

Like Wilber, Hinrichs used map series to portray climatic dynamism, but his maps analyzed the causal agencies shaping the Great Plains climate. Using data from his observers, Hinrichs created a dazzling array of climate maps and map series. Most of Hinrichs's maps were busier and more complex than Wilber's maps. His 1882 map series (fig. 4.4), for example, attested to the existence of local microclimates and to month-to-month variability in rainfall patterns across the state

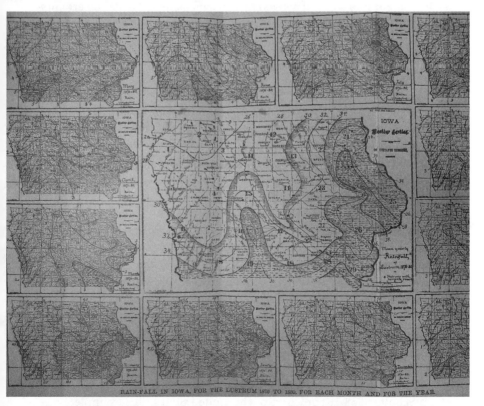

Figure 4.4. "Rain-Fall in Iowa, for the Lustrum 1876 to 1880, for Each Month and for the Year." The map in the center shows mean annual rainfall. The smaller inset maps show the mean for each month. From Hinrichs, *Second Biennial Report of the Central Station of the Iowa Weather Service* (1882).

of Iowa. His two 1877 maps (fig. 3.4 in chapter 3) aimed to establish a correlation between the location of forested areas and average annual rainfall. These two maps are chaotic almost to the point of illegibility; they reflect Hinrichs's belief in the tenuous nature of any scientific claim about correlation or causality.[29] Indeed, Hinrichs admitted that climate theorists still needed to shed "much more light upon this question [of climate change]." Despite his belief in the persistence of uncertainty, Hinrichs claimed that his map series demonstrated the existence of a causal relationship between forest cover and rainfall.[30] Hinrichs believed his map was the first concrete proof that artificial silviculture and afforestation could modify and improve climate patterns.

Wilber and Hinrichs both used their maps to prove society's influence on climate and to further their visions for the future of the Great Plains. The Nebraskan used his maps to legitimize continued railroad building and agricultural settlement; Hinrichs used his maps to highlight the importance of activist forest management.[31] The uncertainty and fluidity of Hinrichs's and Wilber's geographical visions did not prevent them from making confident policy-related claims.

Whose Map Is It Anyway?

Hinrichs and Wilber explained the political and environmental implications of their cartography, but doing so did not guarantee them control over their maps' meaning. Denis Wood has described "the ferocious power of maps to speak for themselves."[32] Sometimes climate maps spoke against their creators. Instead of explaining or fixing climates and environments, many Gilded Age maps inadvertently created new questions, ambiguities, and uncertainties.

Cartographer Charles Schott believed his maps and tables had definitively disproved theories of anthropogenic climate change such as those advanced by Hinrichs and Wilber. During the 1860s and 1870s, Schott worked for two of the key climate research organizations of the late nineteenth century—the Smithsonian Institution and the Department of the Interior. According to historian James Rodger Fleming, Schott used maps in an attempt to "put to rest uninformed speculation" about possible climatic changes caused by settlement.[33] Yet his "Rainfall-Charts" were uncertain enough to allow their reinterpretation by climate change proponents.

Schott's 1868 "Rain Chart of the United States," for example, is replete with unfinished isolines.[34] Like many maps showing average rainfall levels, the 1868 chart uses a combination of lines and shading to show average rainfall. Instead of bounding the entire West within his isolines, Schott allowed some lines to trail off into the surrounding landscape, suggesting either that data was unavailable or that some areas' rainfall levels were too localized and changeable to be portrayed on a small-scale map (figs. 4.5 and 4.6). Tellingly, Schott did not complete the isohyets around the Great Salt Lake. As I discussed in chapter 1, the question of Mormon climate improvement near Salt Lake City emerged as a point of contention as early as the 1860s. Rather than settling the issue, Schott's map left the question of Mormon climatic influences open to interpretation.

Twelve years later, despite the abundance of data that had been collected in the intervening period, Schott created an equally ambiguous map of rainfall levels. Ironically, this map seemed to endorse the theories of Schott's opponents in the climate debate. His 1880 "Rainfall-Chart of the United States" used hachures to depict a small oval portion of Utah as a blue oasis of plentiful rainfall amid a vast arid region (fig. 4.7). The blue hachures indicate that Salt Lake City and parts of its hinterlands, including higher-elevation areas, received more than twenty inches of annual rainfall, a crucial benchmark for agriculture. Throughout the West, high-elevation areas tend to receive more rainfall than surrounding basins and valleys. Mountain microclimate may account for part of the blue island in Schott's map, but the "Rainfall-Chart" does not portray increased rainfall areas in higher-elevation zones elsewhere in Utah Territory. To an ardent expansionist, the map might offer tacit approval of Mormon-based climate improvement theories.

Schott's 1880 map also included a disclaimer reading "precipitation between longitudes 103° and 123° imperfectly known."[35] By acknowledging the uncertainty of much of the western portion of his map, Schott may have facilitated the eventual appropriation of his maps and data by climate change proponents. Just three years after the publication of the "Rainfall-Chart," Henry Allen Hazen cited Schott's work as proof of his argument that the Great Plains and other western regions had seen an increase in rainfall. Hazen, another member of the Washington-based science bureaucracy, asserted that "the gradual increase in rainfall during the past three years is noticeable over a large extent of country."[36] Hazen believed that human influences such as increased silviculture were responsible for much of the recent climate change in the West, and the empty spaces and gaps in Schott's maps allowed Hazen to develop his fluid geographical vision.[37]

Lorin Blodget, a cartographer, "disgruntled clerk," and sometime colleague of Schott's in the Washington bureaucracy, also experienced cartographic appropriation.[38] Whereas Schott's writings opposed theories of anthropogenic climate change, Blodget seemed to vacillate on

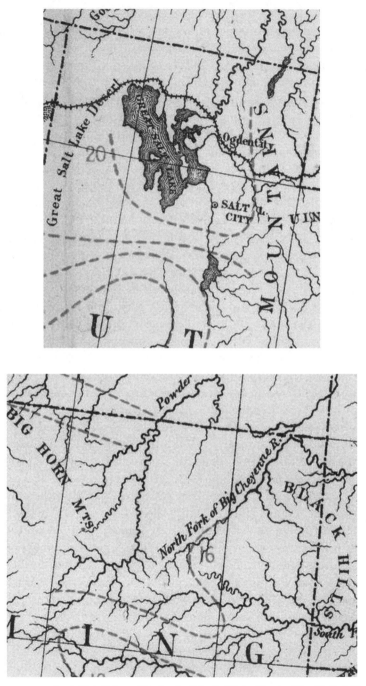

Figures 4.5 and 4.6. Detail from Charles A. Schott, "Rain Chart of the United States Showing by Isohyetal Lines the Distribution of the Mean Annual Precipitation in Rain and Melted Snow" (1868).

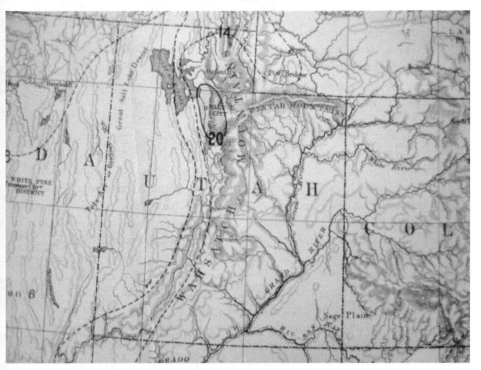

Figure 4.7. Detail from Charles A. Schott, "Rainfall-Chart of the United States Show-ing the Distribution by Isohyetal Curves of the Mean Precipitation in Rain and Melted Snow for the Year" (1880).

the issue. His 1857 study *Climatology of the United States* explained that "the surface condition . . . is a consequence, not a cause" of climate. Early on, Blodget asserted that the extent of forests and agriculture was influenced by climate and could not at the same time influence atmospheric conditions. In 1872, however, Blodget wrote that "the mutual relations of surface to climate cannot be disputed." Blodget came to believe that afforestation could modify local climates, but he remained uncertain whether settlement and the concomitant changes in landscape had caused a large-scale transformation in western weather patterns. During the early 1870s, observation networks like Hinrichs's had yet to proliferate across the prairie. Blodget could obtain only short-term data from a few scattered stations. He believed rainfall records were too sparse "to afford any proper means of determining whether the quantity of rain is greater now than it was ten years since."[39] Judging from his skepticism about available data sources, Blodget did not intend his maps to endorse theories of large-scale climate change. But to some extent his uncertainty may have abetted the reinterpretation of his maps by more strident climate change proponents such as Mark Harrington.

Harrington's 1887 article in the *American Meteorological Journal* used Blodget's maps in an attempt to answer the question of whether rainfall was increasing on the Plains. By comparing Blodget's maps with a later series of maps by Charles Denison, Harrington sought to prove that rainfall levels had increased. His article is unique in that it attempted to estimate the rate of westward movement of cartographic isohyets.

Harrington focused especially on the progression of the twenty-inch rainfall line, a line that many scientists and surveyors believed marked the maximum possible extent of agriculture without irrigation. At the start of his article, Harrington admitted the fickleness of his data and the dangers of making definitive claims about climate change: "The question of increase or decrease of annual rain-fall is a difficult one to decide for any locality. The precipitation in the temperate zone is extremely variable from season to season, and the annual amounts show very great differences." Harrington also alluded to the fact that new weather stations and expansive data-gathering efforts did not always clarify the climate picture: "Of two stations, not very far apart, one may show for many years an apparent average rain-fall higher than that of the other, due to one single storm which passed over the one and not over the other." When discussing maps, by contrast, Harrington used a more confident tone. He concluded that the results were "very consistent in themselves" and showed that isohyets such as the twenty-inch rainfall line were moving West at about "5 miles a year" in some places and "less than a mile a year" in others.[40] Harrington's discussion of maps and isohyets highlights both the cartographic dialectic between uncertainty and certainty and the frequent use of appropriation in Gilded Age debates about geography. In that regard, Harrington would soon get a taste of his own medicine.

A Geography of Limits

Writers who opposed theories of anthropogenic climate change proved as deft at creating uncertainty and reinterpreting maps as climate change proponents like Harrington. Paradoxically, these more skeptical climate theorists used cartographic uncertainty to espouse a fixed geographical vision. They developed a geography of limits—a static imaginary that emphasized the continuity and consistency of western climates and landscapes. Although not by any means radical critics of expansionism, Henry Gannett and Josiah D. Whitney rebuked climate theorists who believed Euro-Americans could remake the map of the West to fit their needs.

Less than one year after Harrington published his article explaining the westward movement of isohyets, Gannett took issue with his claims in a series of articles in *Science*. Gannett, a skilled and influential geographer, doubted the feasibility of grandiose climate modification schemes. He challenged the claims made by supporters of the Timber Culture Act and Arbor Day, describing the silviculture movement as "useless" before asking, "Is it worth while to go on planting trees for their climatic effects?"[41] Gannett countered the geographical vision created by silviculture and agriculture proponents by casting doubt on their maps. In an ironic echo of Harrington, he explained how seemingly objective meteorological data could be deployed "to obtain any pre-arranged result."[42] Gannett declared Harrington's maps and the older Blodget maps to be "at least 99 per cent hypothetical." As for the isohyets and their supposed westward movement, Gannett asserted that "they might as well be drawn a hundred miles on either side of the position assigned to them" by Blodget or Harrington.[43] Despite being a staunch supporter of scientific progress, Gannett understood how modern data collection projects and technologies contributed to the uncertainty of scientific and geographic discourse even as they promised to uncover new facts and certainties.

Prominent surveyor and geologist Josiah D. Whitney shared Gannett's skepticism about isopleth cartography. Whitney voiced his doubts in an 1876 article titled "Plain, Prairie, Forest." Whereas Gannett criticized cartographers who tried to show the westward migration of isolines, Whitney took issue with those who used map series to prove the influence of forests on atmospheric conditions. Whitney did not specifically mention Hinrichs, but he questioned Hinrichs's technique of juxtaposing rainfall maps with forest cover maps. According to Whitney, such comparative methods could prove only correlation, not causation. He believed the West's patchwork of prairie and forest areas was too complex to be explained solely by climate and climatic variation. Pointing to areas in the Plains and Intermountain West that had high rainfall levels but few trees, Whiney asserted that "some other cause than the want of sufficient moisture has operated to prevent this [forest] growth." Being a geologist, Whitney preferred soil-based explanations. He argued that "the character of the soil" and the "distribution of geological formations" bore most of the responsibility for the distribution of the West's plains, prairies, and forests.[44]

Whitney believed his geological insights had major implications

for Euro-American expansion and environmental politics. The quality of the soil, he implied, would impose some limits on the potential
of agricultural development. Whitney derided boosters and scientists
who trumpeted society's power to reclaim arid western lands through
settlement, forest culture, and the resulting climate changes. After ridi-
culing forestry advocates who "are not content unless they can make
their country out to be not only the garden but the arboretum of the
world," Whitney added that major settlement and afforestation initiatives would inevitably encounter geological constraints. As for the
cartographic implications of his analysis, Whitney stated that he had
demonstrated the limits of maps as explanatory tools. "By no amount
of ingenuity," he argued, "can the peculiarities of isothermal or isohyetal lines be made to play with the marked differences in vegetation."[45]

The Limits of Geography

Whitney's doubts about the power of cartography resonated with
other authors who questioned maps' ability to answer pressing questions about expansion and climate change. John P. Finley and Adolphus
Greely, two cartographers and scientists working for the federal government, adopted a much more ambivalent tone than Whitney about
anthropogenic climate change. Neither believed he could definitively
solve the "vexed" question of human influence on climate,[46] yet both
agreed with Whitney about the limits of cartography's power.

Finley, who had a keen interest in tornadoes and worked for the
Weather Bureau (and earlier the Signal Corps), wrote an 1893 government study on the "climatic features of the Two Dakotas." According
to him, isohyets and shading techniques could not portray the most
salient characteristics of the Dakotas' climate. Finley objected to the
quality of available climate data; he complained about the "imperfect
measurement of snowfall and its reduction to rain" while also decrying
the use of "defective apparatus for the measurement of rainfall." Even
with perfect data, however, Finley did not trust the ability of isopleth
mapping to portray and synthesize "seemingly inexplicable variations"
in climate. Finley experimented with other cartographic techniques
such as point symbol cartography. Frustrated by the uncertainty of isohyets, he simply wrote average annual rainfall totals on his maps of the
Dakotas (fig. 4.8). Finley believed his version of point symbol mapping
would better convey the spatial and temporal variability and unpredictability of rainfall patterns, but he admitted that even "this rather novel
and perhaps doubtful method of representation" failed to capture the
complexity of the Dakotas' climate. Still, Finley maintained that his

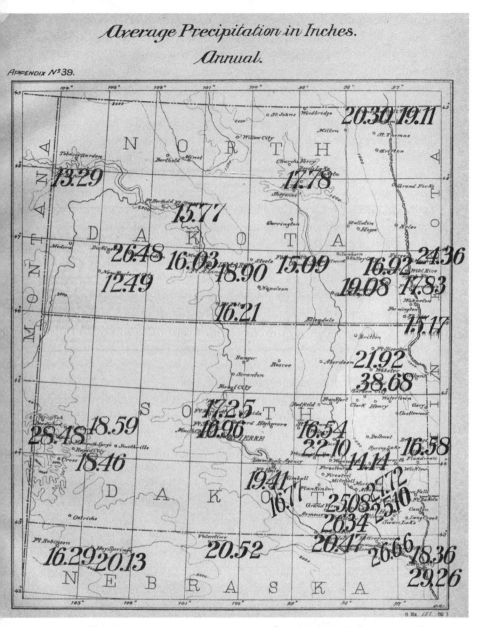

Figure 4.8. "Average Precipitation in Inches. Annual." From John P. Finley, *Certain Climatic Features of the Two Dakotas* (1893).

number-based strategy was preferable to the use of uncertain isolines and "shading of suppositious value."[47] Finley made recourse to seemingly simplistic point symbol maps out of frustration with data-heavy isopleth maps. His work casts doubt on narratives about the inexorable progress of modern cartography toward ever more effective and powerful explanatory techniques.

Like Finley, chief signal officer of the US Signal Corps Adolphus Greely struggled with the limitations of isopleth mapping. In an 1889 report, Greely admitted that rainfall isolines on contemporary maps reflected "personal opinions" as much as anything else. He also stated that maps alone were incapable of proving whether rainfall or general climate patterns followed "any known or definite law." As a high-ranking official, Greely worried about the implications of inadequate maps for the governance of the West. Without more certain cartographic techniques, he wrote, "neither can Congress wisely legislate concerning the varied interests in the Territories, nor can the business enterprise of individuals or corporations safely or economically work out grand results for the arid regions."[48] Ultimately, cartography's promise to facilitate governance and "make sense of the natural world" proved incomplete.[49] Greely's and Finley's writings demonstrate that many cartographers believed in the persistence of uncertainty in the face of novel and sophisticated mapping techniques.

Yet a close reading of late nineteenth-century climate cartography reveals that maps sometimes derived their power precisely from uncertainty. Empty spaces, shifting lines, and blurry boundaries could be just as suggestive as firm cartographic claims. Even as they purported to create a "mirror of nature," mapmakers sometimes announced and admitted their maps' limitations.[50] Figures like Wilber and Hinrichs embraced the paradoxical nature of maps. They did not share Greely's hesitancy about using fluid cartography in land governance or as part of the project of westward expansion. Greely, Wilber, and others created dynamic and uncertain maps to support both their scientific theories and their vision for the future of the American West. The fluid cartographies of the Gilded Age testify to the rhetorical power of geographic uncertainties in the cultural imagination of this purportedly hubristic and unerringly expansionist era.

Rainmakers and Other "Paradoxers"

The expedition's list of supplies reads almost like an order of battle: 16,000 pounds of sulfuric acid, "generators and fittings" for manufacturing 50,000 cubic feet of hydrogen gas, 2,500 pounds of powdered chlorate of potash, 600 pounds of binoxide of manganese, "suitable furnaces and fittings for generating 12,000 cubic feet of oxygen gas," one hundred "strong cloth-covered kites," sixty-eight balloons ten to twelve feet in diameter, as well as sufficient "ingredients for manufacturing several thousand pounds of rackarock powder and other high explosives."[1] "General" Robert Dyrenforth led the expedition.[2] In August 1891 the Dyrenforth party arrived in Midland via the Texas and Pacific Railway and ventured a few miles into the Llano Estacado, the barren "staked plain" of West Texas. The prominent Chicago meat packer Nelson Morris granted Dyrenforth and his men free room and board at his C-Ranch.[3] On the appointed day, half the population of Midland, then a small town of ranchers and cattlemen, joined Dyrenforth's crew on the plateau. They hoped to catch a glimpse of the great spectacle.[4]

A crew of workers assembled and prepared the equipment under the supervision of the general and eight other experts. These eight men represented a cross section of the nation's burgeoning technocratic, bureaucratic, and scientific establishment. In addition to agents from the Department of Agriculture, which funded the expedition, Dyrenforth could count on the support of chemist C. A. O. Rosell, electrician Paul A. Draper, balloonists George E. Casler and Carl E. Myers, and statistician Fred B. Keefer. George E. Curtis, meteorologist for the Smithsonian, was also on hand to explain the scientific rationale behind the experiment.[5] Curtis and most of his contemporaries believed precipitation resulted from the "mingling of different currents" in "the upper strata" of the atmosphere. Dyrenforth and Curtis aimed to encourage some of this mingling in the skies above the Llano Estacado. By creating a series of airborne percussions and explosions, the expedition would also provide the clouds with a steady supply of particulates to help "agglomerate . . . particles of moisture into raindrops." Explosions on the surface would create a "frictional electricity" between the clouds and the surface, further encouraging rainfall.[6]

The men continued their preparations. Some filled the balloons with an explosive mixture of potash and manganese using specially designed piping systems. Others fanned out across the plain, taking their positions among the stunted mesquite trees. Dyrenforth arranged the party into three battle lines, each about two miles long, placed a half-mile apart. Despite his martial demeanor and militaristic approach, Dyrenforth was not solely interested in flagellating the clouds into submission. He also wanted to coax them to yield their treasure of raindrops.[7] Dyrenforth viewed all of nature, especially the atmosphere, as being in a fragile equilibrium. Seemingly insignificant human actions could disrupt this balance. Indeed, Dyrenforth believed humans and their surroundings were inseparably linked. He argued that large crowds or armies could, through their perspiration and breathing, increase atmospheric humidity and lead to greater rainfall.[8]

Finally the men were ready, but Curtis and Dyrenforth urged them to hold their fire until the skies presented an opportunity. As the clouds arranged themselves into a more promising pattern, Dyrenforth gave the signal and the surface-based artillery fired the first salvos. Next were the balloonists: Casler, Myers, and their assistants loosed the spherical balloons into the clouds, waited until they reached the designated altitude, and then ignited them in a near-perfect simulation of thunder. The kites proved less effective. Burdened by a load of explosives, the kites lost their maneuverability and got tangled up with each other, endangering some of the men below. Dyrenforth suspended the kite operation and ordered the artillerymen and balloonists to redouble their efforts.[9]

Rains had been even scarcer than usual in West Texas that year—part of a dry cycle that had started back in the mid-1880s and afflicted much of the High Plains and Southwest. The farmers and ranchers who had ventured into the semiarid shortgrass prairie during the preceding wet years had watched their money, stock, and hope dwindle by the season. Some of these settlers kept encyclopedic records of daily weather patterns in their journals and diaries. Many looked to Dyrenforth and his expedition as potential saviors. And the people of Midland were not the only ones awaiting the results of the experiment; newspapers from as far away as Chicago and New York covered the story, sometimes with lengthy illustrated reports.[10]

The cannonade lasted several hours. Eventually Dyrenforth issued the order and the men stopped bombarding the clouds. Quiet returned to the Staked Plain, but there was still no rain. As the workmen and the experts left their positions and reconvened, a few drops began to fall. The drops turned into a copious rainstorm. In Dyrenforth's words, the water "[transformed] the roadways into rushing torrents and every hollow of the prairie into a small lake." Two full inches of rain! An astonishing amount—more than the average rainfall for the entire month of August in Midland. The following day, reports from cowboys confirmed that the rainstorm precipitated by the Dyrenforth party had extended for hundreds of square miles downwind of the original test site.[11] The journalists raved. "The government rainmakers," wrote one newspaper, succeeded in "breaking a drought of long duration and averting the distress and suffering which would have followed a few more weeks of dry weather." Another paper proclaimed that the rainmakers had "outdone Moses."[12] Dyrenforth conducted two more experiments in the following weeks to prove that the first test was not a fluke. Both succeeded.[13]

* * *

The wave of euphoria generated by Dyrenforth's Midland experiments subsided within weeks. Bitter contention ensued, with many authors

Figure I.1. Note the kite resting behind the two men on the right-hand side of the image. Dyrenforth is the fifth person from the left (seated). Courtesy of the Forest History Society.

Figure I.2. Courtesy of the Forest History Society.

Figure I.3. Courtesy of the Forest History Society.

claiming that the Texas rainstorms would have taken place even without the expedition's efforts. A piece by the Iowa Weather and Crop Service cautioned readers against accepting the results of the experiments "without question or investigation as to the facts." According to the Iowans, a quick glance at government weather maps proved that the rainstorms attributed to Dyrenforth were actually part of a Texas-wide weather event too large to be caused by local bombardment.[14] Adolphus Greely of the US Army Signal Corps depicted Dyrenforth and his associates as opportunists. He took issue with their decision to perform the experiments during August. By working during the few months of the year when rainfall is not uncommon in West Texas, Greely alleged, the rainmakers had showed "the wisdom of serpents."[15] Dyrenforth responded to accusations like Greely's by arguing that "we had a meteorologist in the party who took observations continually, and we made all the experiments at times when there was not the slightest indication of an approaching storm."[16] But the "general" hurt his cause when he attempted to replicate the Midland success in experiments outside El Paso and Corpus Christi and, the following year, near San Antonio as well. Neither attempt conclusively produced rain, giving further ammunition to the growing chorus of skeptics.[17]

The *New York Times* emerged as one of Dyrenforth's harshest critics. Taking a supercilious tone, the paper implied that Dyrenforth was not unlike the confidence men and get-rich schemers populating the late nineteenth-century economy. The *Times* grew even more skeptical of the "General" after Bernhard Fernow of the Division of Forestry, who had always been wary of the Dyrenforth initiative, struck down the rainmaker's claims in a "most serious and solemn manner." Fernow attempted to differentiate his climate change theories from the rainmaker's seemingly outlandish pronouncements. Yet the newspaper interpreted the controversy as a testament to the limits of all climatic and meteorological knowledge, as evidence of the persistence of scientific unknowns: "After all, what do we know about the weather? And, supposing rain did follow Gen. Dyrenforth's efforts, what would it prove? We do not know."[18]

Much of the backlash against Dyrenforth originated from government scientists seeking to insulate themselves from any association with the rainmaker. Dyrenforth had received $9,000 in funding from Congress. As the initial rave reviews of the experiment gave way to skepticism, some critics began to ask why the US government had supported an extravagant and seemingly hopeless venture. In a rebuke of Dyrenforth, academic geographer William Morris Davis wrote that "it is not creditable to congressional action to undertake the experiments upon the artificial production of rain in our present knowledge of meteorology." Davis deemed the scientific rationale behind the Midland tests "a parody of scientific argument." Even as he remarked on the dubious state of contemporary meteorological knowledge, Davis maintained that science needed to be guarded against pretenders like Dyrenforth, figures who preyed on the "sincere belief" of western settlers "predisposed to believe any theory of artificial production of rain."[19]

Perhaps because of admonitions such as Davis's, George E. Curtis, the government meteorologist who participated in the Midland experiment, attempted to distance himself from Dyrenforth. Curtis reportedly supported the project's results in the immediate aftermath of the events in Midland,[20] but he excoriated the "General" in a pair of articles in the *American Meteorological Journal* and *Engineering Magazine*. Curtis attempted to dissociate his federal employers from the Dyrenforth experiment by invoking a legalistic technicality: "The experiments, though made under the direction of the Secretary of Agriculture, were not undertaken on his responsibility nor on his recommendation, but in accordance with the provisions of an appropriation originating in Congress itself."[21] Echoing Davis, Curtis argued that the experiments had undermined "faith in the Government, faith in science, and faith in the honesty and sincerity of the Government's agents." The meteorologist advised politicians and fellow scientists to avoid any further association with rainmaking. "The idea that noises and concussions will produce rain is found to be a part of the folklore of very many primitive peoples," Curtis wrote, "and to revive it now is to reject the light of civilization and to retrograde to a cruder and less rational apprehension of rational phenomena."[22]

Surprisingly, Curtis defended Dyrenforth's findings in the very same articles that labeled rainmaking a primitive and unscientific enterprise. He admitted that the percussion approach could never conjure a storm on a clear day. "It is evident that the experiments have utterly failed to demonstrate that explosions can develop a storm," Curtis wrote in *Engineering Magazine*. At the same time, however, Curtis attributed a definite meteorological effect to the rainmakers' stimuli. Each blast triggered by Dyrenforth's men "was followed by a very noticeable momentary increase of drops." These brief artificial showers remained enigmatic; Curtis speculated that electric stimuli and concussions on the ground may have promoted condensation in the air, but he could not identify a single cause for the rain.[23] Though far too paltry to justify the implementation of large-scale commercial rainmaking, the mysterious showers Curtis described seemed to prove that humans could exert at least some short-term influence on the atmosphere.

Then why did Curtis denounce rainmaking as an affront to science? How could he disavow Dyrenforth when his experiments produced such an "interesting result"?[24] Curtis's seemingly contradictory stance toward Dyrenforth may have been a strategy of self-preservation, rationalizing his participation in the experiments by claiming they had produced some empirical data. Or perhaps, to use the term coined by US Weather Bureau chief Mark W. Harrington, Curtis and Dyrenforth were both "paradoxers." In 1894 Harrington wrote a thorough report on rainmaking that included as much cultural critique as scientific analysis. Writing a few years after the Dyrenforth brouhaha, Harrington did not take the same defensive stance adopted by many scientists in the immediate aftermath of the Midland experiments. Although Harrington described rainmakers as paradoxers, he did not intend to be disparaging. In their own time, Harrington explained, Galileo, Kepler, and Newton also played that role. While "rank paradoxers" were mere charlatans, other more legitimate paradoxers played a crucial role in unsetting the scientific and philosophical orthodoxies of their eras. Since rainmakers created scientific confusion and spurred new questions, Harrington ranked them as "fin de siècle" versions of their esteemed "paradoxer" predecessors. The Weather Bureau chief seemed to grasp that defending science against such figures would be a Sisyphean struggle. With his embrace of paradox, Harrington signaled that his colleagues should accept the unpredictable nature of knowledge production as well as the unintended consequences of empirical experimentation.[25]

Harrington neither denounced nor endorsed Dyrenforth's findings, but he described the general's method as "highly ingenious."[26] The Weather Bureau chief seemed more concerned with the moral quandaries raised by rainmaking than with the outcome of the Dyrenforth experiment or the practicability of specific rainmaking techniques. Anticipating twentieth- and twenty-first-century concerns about geoengineering and cloud seeding, Harrington described the potential risks of rainmaking: "the phenomenon to be produced probably can not be controlled as to area covered, and may occur where it is not wanted." Should the "weather maker prosper," Harrington speculated, "he will often find himself very much embarrassed until our lawmakers have caught up with our advance in the arts, and the volume of the statute books has been materially enlarged."[27]

The late nineteenth-century scientific establishment had an adversarial relationship with rainmakers like Dyrenforth. Technocrats "were long embarrassed about their inability to counter rainmaking."[28] Some experts, such as Fernow, certainly viewed rainmaking experiments as a menace to science. But not all meteorologists and climatologists sought to purge the unresolved mysteries of rainmaking from scientific discourse. Figures such as Harrington and Kansas state meteorologist J. T. Lovewell seriously considered the implications of experimental rainmaking.[29] Harrington and like-minded scientists may have realized they could not control the evolving dialectic between climate change theory and the practice of climate modification. Indeed, rainmaking by government agents and private imitators persisted despite the vociferous objections of Greely, the *New York Times*, and many other critics.[30] Contested climate theories inspired practical experiments; inconclusive experimental results, in turn, prompted more theorizing. This dynamic contributed to the uncertain milieu surrounding questions of human-induced climate change. And rainmakers were far from the only "paradoxers" experimenting with climate change in Gilded Age America.

CHAPTER 5

Mysterious Ecologies

Three Views on Cyclonic Disturbances

For a mysterious figure referred to only as "D.W.W.," antebellum border ruffians and tornadoes had much in common. D.W.W. wrote an article published in the *Atchison (KS) Champion* on May 20, 1885, and picked up the following day by the larger *Topeka Commonwealth*. Alluding to the "Bleeding Kansas" era of the late 1850s, the article drew a link between destructive storms of the Great Plains and pro-slavery thugs. D.W.W. believed the advance of civilization had defeated "cyclone twisters and water spouts" just as it had vanquished pro-slavery forces in an earlier era: "The same enterprising and gallant spirit that drove the ruffian, the guerilla and town-burner from Kansas will also speedily abolish the storm fiend, his natural and brutal ally." He boasted that "in a free, fair, stand-up fight, we have whipped the climate and driven it howling from the field." The newspaper piece focused on "tree growing." D.W.W. viewed the spread of silviculture as a "tremendous revolution" that had enabled Kansans to dominate and improve their volatile climate. According to D.W.W., culture—construed as forest culture and as a concomitant change in general mentality—won the day.[1]

Five years after the *Commonwealth* published the tree-growing piece, Dr. Susanna Way Dodds presented a paper titled "What Causes the Cyclones" at a meeting of the American Forestry Congress held in Washington, DC. Born in Indiana, Dodds practiced medicine in St. Louis, Missouri, where she also served as dean of the Hygienic College of Physicians and Surgeons. An advocate for naturopathy, Dodds is perhaps best known for devising a "hygienic or natural" approach to curing disease that emphasized diet instead of drugs.[2] Dodds's views about bodily equilibrium influenced her environmental ethos—she believed human agency had disrupted the delicate balance among soil, forests, and weather patterns. In contrast to D.W.W.'s militaristic language, she used health-related rhetoric, asking "What . . . is the remedy" for "high wind tornadoes" and other violent weather events? Dodds described the appalling deforestation she had witnessed in Ohio and during her travels along the Pacific coast four years earlier. Her remarks drew on long-standing beliefs about the close relation between health and climate. But Dodds also articulated a nuanced theory of storm formation based on air currents of varying temperatures. She asserted that surfaces and landscapes such as barren, deforested areas could contribute to the formation of "tremendous tornadoes or cyclones." In many parts of the United States, Dodds argued, climate had already been rendered more extreme by human influences: "The April showers are now the exception, and either beating rain-storms or long-continued droughts are the rule." Dodds concluded that clear-cutting and the resulting capricious climate were a "burning shame" that needed to be remedied.[3]

Six years after Dodds delivered her paper, the *Southern Lumberman*, a trade journal catering to the expanding lumbering and forest-products industry, published "Trees and Tornadoes." By issuing a strong defense of the climatic importance of standing timber, the column tapped into the increasingly popular conservationist notion that forest cutting should be carefully managed. The *Southern Lumberman* echoed Dodds's conclusions, affirming her view that trees "maintain parity" in climate. But the journal's August 1, 1896, piece diverged from Dodds in claiming that electrical currents could explain the relationship between forests and extreme weather. Invoking W. B. Hazen of the US Weather Bureau, the journal stated that "a tornado, according to Prof. Hazen, is a violent manifestation of the thunderstorm, and any method of diminishing the electricity of the air will inevitably diminish the tendency to violent outbursts." Since tree growth prevented "intense heating of the soil," the argument went, it would "neutralize atmospheric electricity." At the same time, the *Southern Lumberman* allowed that the precise origin of tornadoes remained mysterious and deserved "a thorough and exhaustive study."[4]

Two Ecologies

Aside from a shared belief in anthropogenic influences on extreme weather events, these three sources seem to have little in common. They range from North to South to West, from a Gilded Age "frontier" ethos to a Progressive Era emphasis on scientific forestry, from shrill boosterism to moralism, from the elite science of Professor Hazen to D.W.W.'s folkloric notions, from a confident faith in society's power to improve nature to a concern with environmental fragility and disturbance. Even the three pieces' intended audiences differed. D.W.W. probably aimed at settlers and prospective settlers, Dodds at fellow forestry advocates, and the *Southern Lumberman* at industry experts or perhaps at members of the emergent field of scientific forest management.

Despite their differences, however, the three pieces shared an ecological sensibility: a belief in the interdependence and interconnections among soil, humans, water, vegetation, nonhuman animals, and climate. Dodds, for example, believed that disrupting the equilibrium between forests and the atmosphere would produce a series of concatenated negative effects such as the drying of watercourses, the disappearance of birds, and the proliferation of harmful insects. D.W.W., on the other hand, believed that forest culture on the Great Plains had triggered a virtuous ecological cycle. Newly created woodlands provided a refuge for wild animals and stock while also creating "millions of spontaneous springs, brooks, and creeks." Though the *Southern Lumberman* did not address broader ecological questions, it depicted terrestrial and atmospheric conditions as being in an intertwined, dialectical relationship.[5]

The German polymath Ernst Haeckel coined the term "ecology" in 1866, and ecology coalesced as a discipline over the course of the twentieth century, but locating the origins of ecological thinking is difficult.[6] As Donald Worster and many other scholars have observed, naturalists and other environmental thinkers from the nineteenth century and earlier developed an eclectic range of ecological paradigms. Defining ecology is as challenging as tracing its origins. Worster's *Nature's Economy* describes ecology as a "point of view that sought to describe all of the living organisms on earth as an interacting whole."[7] Many early ecological theorists included abiotic components as well; they explored the interrelations and interdependence among organisms and their inanimate surroundings.

Late nineteenth-century Euro-Americans such as D.W.W., Susanna Dodds, and the editors of the *Southern Lumberman* certainly did not invent the notion of ecological interconnections. Their views reflect the influence of some of the main currents shaping nineteenth-century environmental thought. By characterizing ecological connections as elusive, the forestry publication followed in the footsteps of the renowned explorer Alexander von Humboldt, who, in the words of historian Aaron Sachs, "referred to ecological links as 'occult forces' because they work in such mysterious ways." Dodds's and D.W.W.'s writings show the influence of George Perkins Marsh, a conservationist who understood nature in terms of equilibrium and disturbance. All three cyclone essays also evoke a strain of environmental holism that predated Marsh's seminal 1864 work *Man and Nature*, a tradition best described by Conevery

Valenčius in her study of settlement in antebellum Missouri and Arkansas.[8] Dodds, D.W.W., the *Southern Lumberman*, and many contemporaries drew from these antecedents and devised an idiosyncratic blend of paradoxical ecological notions.

Though Gilded Age forestry advocates did not develop a single cohesive environmental theory, they shared a belief in an uncertain ecology.[9] They understood forests, society, and weather patterns as a complex, interconnected network whose inner workings remained mysterious and unknown. Their scientific treatises, articles, and pamphlets reveal an assemblage (or ecology) of connected notions about the relations among trees, people, climate, and other living or abiotic agents. Uncertain ecology ranged across spatial and temporal scales; it included holistic or humoral theories but also mechanistic natural relationships, some within the reach of human knowledge and human agency and some beyond. Despite the indeterminacy of their ecological paradigm, figures such as Bernhard Fernow and Franklin Hough used uncertain ecology as the basis for their forest planting and conservation initiatives. Though not removed from the forces of conquest and dispossession that shaped Gilded Age culture, late nineteenth-century foresters used climatic ecology as the basis for an environmental vision rooted in humble appreciation of the interconnections between humanity and the vast scale and mystery of the nonhuman world.

But this story is not about a few environmental visionaries ahead of their time and forgotten by their successors. Nor is it an account of some amateur proponents of folkloric "forest culture" destined to be eclipsed by modern forestry, ecology, and science.[10] Uncertain ecology pervaded Gilded Age culture and persisted into the early twentieth century. Although they were not shared by all foresters and forestry advocates, notions of ecological uncertainty and complexity suffused the work of relatively famous figures such as Fernow and the writings of more obscure forestry advocates. A few prominent foresters certainly played a key role in popularizing climate-based forest advocacy. At the same time, however, Fernow, Hough, and their colleagues belonged to an epistemological ecology that blurred the boundary between high and low culture, popular and elite science. While emphasizing a few leading exponents of uncertain ecology, I also hope to convey that these environmental theorists belonged to a sprawling network of mutually influencing thinkers. Late nineteenth-century Euro-American forestry advocates participated in parallel ecologies: a complex system of knowledge creation and an intricate network of forces shaping climate change and climate stasis.

John Warder: "Genius of the Tongueless Mysteries"

After John A. Warder died on July 14, 1883, at age seventy-two, eulogies and elegies poured in from his fellow arboreal enthusiasts. One obituary listed Warder's works, including *American Pomology* and *Hedges and Evergreens*, and described him as a person "of vigorous intellect and keen perceptions" who "cultivated" wide-ranging interests and "devoted his long life, unselfishly, to the diffusion of information on his favorite studies." Another writer, J. N. Matthews, crafted a Romantic tribute to Warder:

> His was the gentle spirit of the woods,
> The genius of the tongueless mysteries
> Eternally that dwells within the trees,
> The flowers, the grasses, and the bursting buds:
> A member of their secret brotherhoods,
> He caught the everlasting sympathies
> Of all the lute-lipped leaves. He held the keys
> Of nature's variant moods and solitudes.
> A druid gray, his loving life-blood leapt
> In transport tremulous, beneath the power
> Of beauty and of symmetry that slept
> Within the petals of the frailest flower.
> Noblest of all the songless bards, he kept
> His great soul stainless in his Eden-bower.[11]

For Gilded Age silviculturists like Matthews, nature's mystery held a strong allure. The author of the poem valued Warder's closeness to the inscrutable ecological workings of the forest. Considering the emotive, almost saccharine, quality of writings like Matthews's, it is tempting to depict Warder and other proponents of forest culture as the last exponents of an ascientific environmental tradition soon to be displaced by "forest management," a more rigorous discipline grounded in notions of efficiency and progress.[12] Yet Warder's eclectic and seemingly contradictory writings bridged the divide between impressionistic forest culture and rationalist forestry. His brand of forest advocacy reveals that uncertain ecology proliferated across a broad range of Gilded Age intellectual currents.

Warder developed an interest in trees and horticulture long before the professionalization of forestry. Born in Philadelphia, he met noted naturalists John James Audubon, André François Michaux, and Thomas Nutall before moving with his family to a farm in Ohio, where he "developed an intimate knowledge of agriculture and fruit-growing."[13] Warder also devoted himself to studying astronomy and meteorology but, like many nineteenth-century polymaths, settled on medicine as his primary career. He worked as a doctor in Cincinnati while also publishing numerous articles in horticulture journals and serving as a member of the Ohio State Board of Agriculture. During the 1870s Warder emerged at the forefront of the growing forest advocacy and forest protection movement. Motivated by fears of a looming "timber famine" and by faith in the innumerable benefits of tree planting, Warder and like-minded environmental thinkers worked to spread the doctrines of forest conservation.[14]

Warder attended the 1873 International Exhibition in Vienna and gained further prominence in the forestry movement after publishing a report based on his discussions with European forestry experts at the exhibition. His report juxtaposed destructive American attitudes toward trees with European forestry, an "art which has grown to great perfection under centuries of nursing." The collaborative spirit of the Old World foresters also impressed Warder. Europeans, he reported, "found it advisable to hold international congresses of the forest-managers of different countries, in order to exchange views of practice, to discuss principles, and to suggest regulations for their respective governments."[15] Soon after his return from Vienna, Warder founded the American Forestry Association, probably the first organization devoted to the cause of forest culture in the United States. The American Forestry Congress, another newly formed group of promoters and advocates, merged with the American Forestry Association in 1882, forming a sweeping national-scale organization that held annual and semi-annual meetings and published its proceedings.

Warder and his fellow members of the American Forestry Association had two related aims: to protect existing forests from clear-cutting and to establish forest-planting initiatives in denuded areas.[16] To accomplish these goals, forestry advocates sought to sway politicians and prominent figures while also instilling an appreciation for forests among farmers, settlers, and other Euro-Americans working on the land. Climate theory played a central role in the efforts of Warder and his allies. Though published a few years after Warder's death, the 1885 *Proceedings of the American Forestry Congress* highlights the importance of climate change in forestry advocates' environmental vision. The congress's organizers listed climate as the first of a series of "considerations" for convening their annual gathering. They asserted that "the general and local climatological influence of forest areas, though not yet clearly defined and numerically demonstrated, is beyond doubt established by historical and experimental evidence."[17] Similarly, Warder himself admitted that the complex ecological mechanisms at the root of climate change remained mysterious even as he cited climatic improvement as a major benefit of tree planting. "Time," he wrote, "may be necessary to eliminate the possible errors arising from cycles dependent upon cosmical causes not yet fully understood; but let us have credit, and let the judicious plantations of trees have the credit for their influence in modifying the local climate of the farms, townships, counties, and states." Warder sought recognition for his cause's potential in the face of long-term and large-scale scientific uncertainties.[18]

In speeches and papers during the late 1870s and early 1880s, Warder struggled to reconcile his belief in ecological inscrutability with his urgent message that "tree-planting is indeed a necessity." At times he relegated complex climatic causality to the background in order to craft a more straightforward and immediate message. He told an audience of Nebraskans that "the great question of general climatic influence ... need not concern us" while urging them to "meliorate the conditions of our own immediate surroundings by planting groves, shelter-belts and forests."[19] Warder delivered the Nebraska speech in 1878; four years later he published an article claiming that "no one can any longer question the validity of the claims that forests do modify the climate." In the same piece, Warder seemed to adopt a straightforward ideology of environmental improvement, the same rhetoric that often featured in expansionist treatises. Nature, he claimed, needed Euro-American intervention to attain its full potential: "Planting directed by human brains is better, and the results will be more satisfactory, than trusting to natural reproduction, for it enables us to do work more thoroughly, more evenly and more judiciously."[20] Warder's emphasis on positivism and efficiency prefigured the approach of foresters from the 1890s and the early twentieth century. According to historian Nancy Langston, these Progressive Era experts "developed an overreliance on universal scientific theories that made it difficult for them to value complexity, inefficiency, uncertainty and redundancy."[21]

Yet Warder did not abandon uncertain ecology. Even as he praised the virtues of "human brains," he also exhorted his listeners: "In our lamentable ignorance—let us follow nature's plan! Examine the sandbars on the rivers, prepared and planted by the wind and wave. Look at the barrens and the bluffs that have been planted by the squirrels and the gophers and the birds, in addition to the air and wind—these are all instances worthy of our study."[22] Only by considering the complex interconnected workings of nature would Euro-Americans succeed in spreading forest culture and, eventually, in ameliorating their climate as well. Warder's writings encompassed both rationalist and uncertain ecologies, showing that perhaps the "gospel of efficiency" and the gospel of mystery were not mutually exclusive.[23]

In addition to the tension between uncertainty and proto-Progressive efficiency, other seeming contradictions marked Warder's forest advocacy. Warder was an amateur who railed against amateurism. At times he supported a bottom-up approach to conservation, one rooted in the "earnest and practical knowledge" of tree planters working on the land. But in other instances Warder stressed the importance of professional expertise and urged silviculturists to abandon piecemeal planting efforts in favor of a planned initiative under the aegis of a group of trained scientists.[24]

Warder also vacillated in his views about the relation between forest conservation and capital. In an 1882 speech before the American Forestry Congress annual meeting in Montreal, he argued that "the long rotation of most trees puts the profits of the harvest beyond a generation of men, hence, they who plant can rarely expect to reap." Warder's speech implied that spreading forest culture would necessitate rethinking the short-term profit motive that characterized much Gilded Age economic expansion.[25] During the same year, Warder issued a statement in favor of expanding the Timber Culture Act to include a broader range of eligible tree types. In this piece he fused economic and climatic considerations, arguing that afforestation of previously treeless areas in the Great Plains would "check the force of the winds," "make these broad expanses habitable," and "meliorate the climate" while also providing fuel and building materials for railroads. Warder endorsed the alliance between government and private enterprise—primarily railroad corporations—for its role in developing the West and disseminating silviculture.[26] Although he acknowledged that early railroad-sponsored experimental plantations were "spasmodic efforts . . . abandoned during the financial panics," he also praised the Union Pacific and Northern Pacific Railroads for establishing successful plantations on the Plains.[27] Clearly, conflicting currents of environmental thought and political economy underscored Warder's forestry writings. Similar contradictions and schisms marked the articles, papers, and speeches of Warder's fellow silviculture proponents and American Forestry Association members. The political and cultural implications of uncertain ecology remained contested even among authors who agreed about the importance of conservation and climate improvement.

Over the 1870s and 1880s, debates about the veracity of climate change theses raged among Euro-American environmental thinkers and scientists. Within the forest culture movement, however, authors tended to agree that society could influence weather patterns through agriculture, afforestation, and deforestation. Most forestry advocates also shared the basic tenets of uncertain ecology: that humans, plants, and other beings formed a complex, interconnected network; that the inner workings of this complex network remained mysterious; and that climate change was a symptom of the relative health or imbalance of the natural system.

Few documents better reflect this ecological paradigm than the first volume of the *American Journal of Forestry* (1882–83), a publication containing papers by John Warder and many of his closest colleagues. An essay by Benjamin Gott of Arkona, Ontario, encapsulated the foresters' doctrine: "How intimately close is the relationship that exists between the departments of the natural world, between the vegetable and the animal kingdoms, between the merest vegetable and the highly organized beauty of the air." Although Gott knew natural relationships to be "intimate," he did not purport to know their inner workings, asking, "When shall we learn of the proper relationship of one part to another in the arrangement of nature?" Like many of his colleagues, Gott did not believe that ecological unknowns should prevent people from taking up the cause of forest protection and tree planting. He described a mysterious dialectic between forests and the air and followed it with a succinct command: "The action of the trees and the reaction of the atmosphere is constantly going on, and every time man receives blessings by the mysterious arrangement. Plant trees for moisture." For Gott, tree planting served as an exercise in humility about the limits of human knowledge and about the minute scale of society before the "mysterious arrangement" of nature: "Let us each in our humble way strive to add our humble mite to the sum total of our engagements of this humble life . . . by planting a few trees to live and testify [to] us after our heads are laid low and our hands are still in everlasting rest." Gott's paper highlights a paradoxical aspect of uncertain ecology. He viewed tree planting as an act of environmental humility that, if carried out on a sufficient scale, would allow humans to improve and perhaps master the vagaries of climate.[28]

Many other papers in the *Journal of Forestry* shared Gott's views on forests and climate, and especially his belief that trees "correct the extremes of our seasons." Leonard B. Hodges, for example, echoed Warder's assertion that windbreaks—strategically arrayed rows of planted trees—could render climatic conditions more equable. Hodges explained how windbreaks would allow Great Plains farmers to mitigate the force of blizzards.[29] H. M. Thompson's contribution to the

Journal also stressed the importance of extreme weather alleviation in the Plains. Yet Thompson, a resident of Lake Preston, Dakota Territory, diverged from Gott and Hodges on a crucial point. Whereas many forestry advocates seemed to endorse any and all forms of tree planting, Thompson warned that haphazard, capital-intensive afforestation efforts would have a negative influence on climate. He contended that the "form of planting generally in vogue, for the purpose of growing timber for economic use" can result in "sudden changes in temperature, and in such disturbance of electrical and atmospherical currents as tending to increased occurrences of tornadoes and cyclones."[30]

The tensions within Warder's work manifested themselves in disagreements among his disciples. Some viewed forest planting as a way of making a profit while improving climate; others, such as Thompson, developed a radical critique of capital-oriented afforestation. Some farmed themselves or viewed farmers as key allies in planting efforts, while others saw farmers as the scourge of forests, as backward settlers hell-bent on clear-cutting. Some believed afforestation should be left to private enterprise, others, to the federal government. The lack of cohesion among the authors of the *Journal* indicates that even though uncertain ecology shaped the forestry movement, it could not unify the eclectic voices and interests involved. In recent years Bruno Latour has called for the creation of a "politics of nature" rooted in "uncertainty about the relations whose unintended consequences threaten to disrupt all orderings, all plans, all impacts." Perhaps the work of Gilded Age foresters highlights the challenges inherent in creating a political ecology of uncertainty.[31]

And yet the notions of complexity and uncertainty did provide some common ground. They fed both utopian dreams and visions of catastrophe. From the perspective of many forestry advocates, interfering with a seemingly small component of a vast and mysterious natural system could inadvertently trigger an ecological or climatic disaster. For writers like Gott, the inscrutable workings of nature offered a fleeting glimpse into happier futures. Whether conceived in utopian or dystopian terms (or both), uncertain ecology gave silviculture proponents a shared sense of urgency.[32] Additionally, the Forestry Association, the *Journal of Forestry*, and other forums for debating natural mysteries brought together famous statesmen and obscure characters such as Gott. They prompted Euro-Americans from across the political spectrum to collaborate.in their efforts to promote forest culture. To some extent, they flattened distinctions between academics, bureaucrats, farmers, citizen-scientists, and politicians.

The prominent Kentucky politician Cassius M. Clay, sometime abolitionist and, later, ambassador to Russia, appeared in the pages of the *Journal of Forestry* alongside Hodges and Thompson. In his contribution to the *Journal,* Clay issued a categorical statement about the healthful properties of forests—"all trees destroy malaria"—before advancing an idiosyncratic theory of forest-based climate change. Clay believed that forests influenced climate only in areas where the primary cause of rainfall was the confluence of warm and cool air currents. Weather patterns in the West, where, according to Clay, large-scale landforms like mountain ranges determined the extent of rainfall, would neither gain nor suffer from deforestation or afforestation. "The Lion of White Hall" insinuated that forestry advocates should refocus their efforts on older states suffering from clear-cutting instead of embarking on quixotic planting efforts in the arid and semiarid West.[33] Alongside the schism between easterners and westerners, Clay's article brings to light another unresolved tension among forestry advocates. Many of Warder's disciples supported afforestation, reforestation, and the preservation of existing forests, but some argued that only one kind of initiative should take priority.

Forestry advocates proposed myriad causal mechanisms and ecological relationships as explanations for climate change, ranging from Clay's shifting air currents to Hazen's electrical forces. The forestry movement's emphasis on environmental interdependence and interconnections precluded privileging a single causal explanation for climatic changes. Similarly, the sprawling and diffuse character of Warder's associations prevented a single ecological ideologue from imposing their views on the movement. In her study of early twentieth-century colonial entomologists and epidemiologists, Helen Tilley shows how in some cases experts' adherence to "ecologies of complexity" could "temper their utopian visions."[34] The influence of complex ecology on the silviculture movement was even more ambiguous: it gave rise to utopian dreams and environmental anxieties, creating common ground while also engendering schisms and contestation.

"Undeviating and Perpetual": Julius Sterling Morton and Arbor Day

In contrast to Cassius M. Clay, who occupied the threshold between pragmatism and opportunism, Julius Sterling Morton was nothing if not consistent. He was a dyed-in-the-wool Democrat who espoused a different brand of expansionism than most Gilded Age climatic and environmental theorists. Morton supported the expansion of slavery in the 1850s and later proved more willing to criticize yeoman agrarianism than many of his Republican contemporaries.[35] Despite the political tensions underscoring forestry debates, Morton did not hesitate to advocate for forest culture alongside Republicans such as Nebraska governor Robert Furnas, a political rival.[36] Morton participated in the same broad epistemological ecology as Warder and his followers. He tended to see eye to eye with the Ohio doctor, as evinced by his "most eulogistic and flattering" remarks introducing Warder to a Nebraska City audience in 1878.[37]

Morton's family eventually earned renown for its salt company, but back in the late nineteenth century, Julius Sterling Morton derived much of his fame from having founded Arbor Day. Born in Adams, New York, in 1832, he spent much of his youth in Michigan and studied at the University of Michigan. After brief stints as a journalist for the *Chicago Times* and *Detroit Free Press*, he moved to Nebraska City in 1854. He edited the *Nebraska City News* and dedicated himself to horticultural and agricultural experiments on his 160-acre farm. As a Democrat in a predominantly Republican state, he struggled to win elective office and failed in his repeated efforts to win the governorship after Nebraska attained statehood in 1867. In the 1890s, President Grover Cleveland appointed Morton secretary of agriculture for his agricultural knowledge and his loyalty to the Democratic Party.[38] Morton dedicated himself to the cause of forest culture through the ups and downs of his political career, earning a place alongside Warder as one of the nation's foremost silviculture advocates. Aside from some cursory credits for founding Arbor Day, Morton has received little attention from scholars.[39]

Morton first broached the topic of a holiday dedicated to forest culture at a January 4, 1872, meeting of the Nebraska State Board of Agriculture. Furnas, who wrote a laudatory 1888 history of Arbor Day, noted that Morton's resolution to establish the celebration "was unanimously adopted, after some little debate as to the name, some present contending for the term 'Sylvan' instead of 'Arbor.'" Arbor Day's proponents settled on April 10 as the date "to be set apart and consecrated for tree planting in the state of Nebraska." A broader, year-round Arbor Day movement coalesced among supporters of the holiday. Numerous factors spurred the rapid growth of Arbor Day in Nebraska, ranging from local boosters' desire to fulfill Timber Culture claims to the alienating effect of treeless landscapes on recent emigrants from the East.[40] Morton himself likened the experience of encountering treeless landscapes to falling ill.[41]

Especially early on, advocating Arbor Day was a didactic and expansionist effort. Morton and his collaborators sought to educate Euro-Americans, especially prospective settlers and schoolchildren,

about the "meaning of their planted tree with regard to the development of the nation."[42] As B. G. Northrop explained in an 1885 piece,
the early history of Arbor Day was inextricably entwined with notions
of railroad and settlement-driven climatic improvement. "By pen and
tongue," Northrop claimed, Arbor Day supporters had proved that the
"immense plains of the new West" could be "made habitable and hospi-
table by cultivation and tree-planting."[43] Morton extolled the beneficial
climatic properties of trees in his Arbor Day pamphlets and proclamations. The prospect of climate change gave the Arbor Day movement
a redemptive and utopian edge: tree planting would allow settlers on
the Great Plains to atone for the environmental and climatic sins of
the East. While Nebraskans created an ideal "civilization" on the Plains
through forest culture, Easterners continued to "hasten inevitably and
resistlessly [*sic*] the calamitous end of the woodlands," an outcome that
would result in "long-continuing droughts and cyclones careening over
the shorn earth, each year with more frequency and more and more
destructive force."[44]

Morton delivered those remarks at the 1885 meeting of the American Forestry Congress in Boston. Nebraskans like Morton and Furnas sometimes trumpeted the Arbor Day movement as a singular and
unique effort, but theirs was one of many contemporaneous thrusts in
favor of forest planting and conservation. As Arbor Day spread from
Nebraska to the rest of the United States, it lost some of its western
character and fused with the broader forest culture movement.[45] Still,
Morton and his collaborators tended to emphasize afforestation and
new plantings over preservation and reforestation, ensuring that Arbor
Day literature retained some of its western flavor.

A close examination of Morton's writings reveals that western Arbor
Day proponents shared Warder's and his colleagues' views on ecological uncertainty. The Nebraskan invoked the "occult chemistry of nature," a phenomenon "as wonderful in mystery as the depths of eternity
itself." Morton took his ecological paradigm one step beyond Warder's,
going further in his efforts to portray humans as part of a complex and
inscrutable natural system. Morton's 1885 piece "Arbor Day" outlined
his radical ecological vision. It presented silviculture as a secular religion, claiming that "in no form of belief" can "be found a ceremonial
which vitalizes faith as does the act of tree planting." Echoing Warder
and Gott, Morton wrote that "the inter-dependence of animal and
vegetable life is undeviating and perpetual." Perhaps in an effort to re-
assure readers steeped in Judeo-Christian ideology, Morton gestured
toward the anthropocentric logic of conquest: "We declare the animal
kingdom to be superior to the vegetable, and proclaim man emperor of

both." But using a millennial time scale, Morton went on to show that society's victory over the "vegetable kingdom" was illusory. "Time at last tells the real truth," he claimed, and the passage of "eons" would reveal both humanity's complete reliance on vegetation and the inseparability of humans from nonhuman nature. Morton drove home his point by speculating about the dystopian consequences of crop failure: "So dependent is man upon plants, foliage, and fruit that . . . the skipping of a single year of plant life, would turn from life into death every animal organism on the globe." Morton went so far as to question the "physical individualism" of every human, telling his readers that "every muscle, every fiber, and tissue in these hands, in your hands, was once animate in plant form."[46]

If Morton's willingness to challenge the dichotomy between humans and nonhumans set him apart from some other exponents of forest culture, it was not unique in nineteenth-century Euro-American environmental thinking. Historian Linda Nash has observed that, for many health-minded nineteenth-century writers, endemic diseases and miasmas broke down the "assumed separation" between society and nature, between environments and individuals. Morton and many Euro-American settlers perceived the body as permeable. The ephemeral boundary between humans and their surroundings added another component to uncertain ecology and, according to Nash, gave rise to anxieties about settling new landscapes. Like Arbor Day rhetoric, nineteenth-century medical discourse "asserted the agency of white colonizers while acknowledging their vulnerability to the complex agencies of nature they did not fully understand."[47] Morton identified the tension Nash described and reveled in it, proclaiming that "nature teaches by antithesis." While he viewed ecology as mysterious, Morton also believed he had grasped the dialectical workings of nature. The course of westward expansionism, he claimed, had been a process of thesis and antithesis orchestrated by a higher force. After allowing Euro-Americans to destroy eastern forests, nature "unfolded to the great vision of the pioneer" the treeless plains "as a great lesson to teach him . . . the indispensability of woodlands."[48] Uncertain ecology served to rationalize Morton's unapologetic ideology of expansionism.

In contrast to some members of the American Forestry Congress, Morton believed that for all its mystery and "lessons" about humility, nature's plan would inevitably lead to progress and improvement for Euro-Americans.[49] Morton's sense of inevitability extended to his views on Native Americans, whom he viewed as a people "who must die, and a few years hence only be known through their history as it was recorded by the Anglo-Saxon."[50] Uncertain ecological sensibilities sometimes dovetailed with racist expansionism. Especially in the Great Plains, Arbor Day celebrations rarely prompted their participants to question the logic of development and conquest.

As Arbor Day spread, however, it acquired many new meanings, not all of them congruent with the ideas of the movement's founder. Pamphlets and "programmes" for "Arbor Day Observance" shed light on the peculiar arboreal culture that formed around the observance of Arbor Day. An 1893 program for Arbor Day celebrations included both a "forest hymn" and a rendition of "The Star-Spangled Banner," indicating that Arbor Day rituals contained aspects of both religious and civic ceremonies.[51] John Peaslee's 1884 "Exercises and Directions for the Celebration of Arbor Day" reveals that communal tree planting proceeded in a methodical and orderly manner (fig. 5.1). Peaslee was from Cincinnati, and in Ohio, Arbor Day ceremonies often included naming newly planted trees for deceased people. In a letter to Peaslee, politician Samuel F. Cary praised the naming practice in glowing terms: "Imparting to waste spaces more than their pristine beauty and associating the names of departed ones with our work is a poetic and sublime concep-

Figure 5.1. "The Planting of Presidential Memorial Trees in Eden Park," Cincinnati, April 27, 1882 (the ceremony likely memorialized President James A. Garfield, who had been killed the preceding year). Image at top left depicts "Arbor Day Planting Trees." These Arbor Day celebrations were organized by the American Forestry Congress. From *Frank Leslie's Illustrated Newspaper*, May 13, 1882. Courtesy of the Forest History Society.

tion. It symbolizes our faith in a resurrection to a higher and better life when the hard struggles of this sin-cursed world are over."[52]

For Peaslee, naming trees demonstrated that Arbor Day had moved beyond its origins as an expression of Nebraska pioneer hubris. He recommended a rigorous program of "literary exercises" to accompany the "beautiful custom of planting trees." The Ohioan viewed Arbor Day as an opportunity to educate the "children and the public at large" about the "great importance to the climate" of forest planting. Like Morton's version of Arbor Day, Peaslee's emphasized the humble scale of humans relative to "nature" and the crucial role of forests as regulating influences in a complex ecological and atmospheric system. But whereas the Nebraskan maintained his belief in the inevitability of cultural and climatic progress, Peaslee stressed contingency: "Shall the future of this great republic be made uncertain by a gradual deterioration of soil and climate, or shall it forever remain the happy and comfortable home of the free?" Peaslee sought to dissociate Arbor Day from laissez-faire capitalism and profit motives. He viewed the early history of the movement in Nebraska as corrupted by its emphasis on "planting trees for economic purposes" and proudly claimed that the "celebration of 'Arbor Day' by planting memorial trees with literary and other exercises" originated in his hometown of Cincinnati.[53] For Peaslee, carelessness was an intrinsic component of pioneer attitudes, a symptom of deep-rooted elements of American culture and capitalism that needed to be expunged from the national psyche.

Many other forestry advocates shared the Ohioan's disdain for early settlers. In *Arbor Day*, Furnas quoted forester Bernhard Fernow as saying that Arbor Day marked the end of the "era of forest destroyers" and the beginning of a "new era in American life—the era of forest planters."[54] Perhaps some of Arbor Day proponents' contempt for early settlers originated in their elitism. The didactic aspect of Arbor Day literature contains a classist tendency. Educated and relatively affluent authors such as Morton and Peaslee believed it was their duty to instill a respect for trees and a conservationist ethos among the teeming masses of tree destroyers. In contrast to the writings produced by Warder and his forestry associations, which often mentioned the need to learn from the practical experiences of hardscrabble farmers, Arbor Day publications rarely included the voices of working-class Euro-Americans. Indeed, the condescending tone of some Arbor Day proponents likely alienated their audiences.[55]

Despite its occasional elitism, Arbor Day rhetoric challenges the notion that all nineteenth-century afforestation initiatives originated from pecuniary and expansionist imperatives. Certainly, many late nineteenth-century forestry advocates had an economic or ideological interest in tree planting.[56] And expansionists like Morton used forest culture to justify exterminationist attitudes toward Native Americans. But the unresolved struggle over the meaning of Arbor Day indicates that uncertain ecology, climate theory, and forest advocacy cannot be dismissed as simple justifications for conquest.

A "Concert of Observation": Franklin B. Hough's Statistical Ecologies

Morton and most of his fellow Arbor Day proponents often fore-grounded climate in their writings. Ultimately, however, they privi-leged forest advocacy over pursuing climate knowledge. They cited meteorological improvements, including the decreased frequency of extreme storms, alongside countless other benefits resulting from tree planting. Franklin B. Hough, by contrast, dedicated himself to advanc-ing climate science while also supporting forest culture, planting, and conservation. A longtime resident of the state of New York and a man of considerable energy—he once stated, "I seek repose in labor"— Hough paired his belief in uncertain ecology with a veritable obses-sion with data. In addition to collecting his own climatic statistics, he sought to glean meteorological and environmental information from farmers, tree planters, and railroad men throughout the United States. Like Morton and Warder, Hough described humans, trees, and climate as mutually influencing elements within a mysterious and fragile ar-rangement. He viewed violent weather events and "the growing ten-dency to floods and drought" as "common knowledge" and as proof that the stability of enigmatic natural systems had been jeopardized by human agency.[57] But whereas Warder sought "credit" for the cause of forestry in the face of uncertainty, Hough endeavored to build an epis-temological ecology—an intricate system of data collection that would uncover ecological mysteries and give rise to a new, more participatory brand of forest advocacy.

Hough began collecting meteorological and agricultural data on his own during the 1840s, first as a student and then as a medical doctor and an amateur scientist. By the 1870s, he was working as a federal agent for the US Department of Agriculture and oversaw a variety of data collection initiatives. In his capacity as a state and federal statistician, Hough solicited agricultural, climatic, and tree-related data by send-ing a variety of "circulars" (or questionnaires) to potential respondents from a wide range of classes and professions. Hough's methods elicited rebuke from at least one critic, Thomas Meehan, who wrote in an 1873 piece that "Dr. Hough has never been engaged practically in making earth observation" and that his claims "depend for their accuracy on mere newspaper rumor." Despite being a proponent of forestry, Meehan did not believe in climate change either as a scientific thesis or as a rationale for conserving or preserving forests: "A good cause is never aided by bad arguments."[58] Hough may have encountered classist resistance to his inclusive approach to data—perhaps Meehan suspected his use of nonexpert observers. But Hough persevered, maintaining his long-standing belief that "a continued and intelligent concert of observation" would reveal the "peculiarities" at the heart of "artificial" climate changes.[59]

Hough's writings reflect often-repeated narratives about the mid-to-late nineteenth century: the bureaucratization and professionalization of disciplines such as forestry and climate science; the shift from anecdotal, folkloric belief to modern data-based science; and, more broadly, the effort to collect information and rationalize environmental knowledge in order to further state governance and control.[60] At the same time, however, Hough underscores the tensions and false dawns that marked the beginnings of American scientific and cultural modernity. His data collection initiatives, for instance, cataloged the environmental damages caused by expansion and development.[61] Hough grasped the paradoxical nature of modernizing projects such as his data collection. Instead of uncovering ecological laws, he raised new scientific questions and doubts about the "waste-land" left in the wake of expansion and development.[62] Instead of rationalizing and systematizing information flows, his democratic approach to data created a confounding trove of numbers and observations. For Hough, uncertain ecology proved to be a receding horizon. He worked tirelessly until his death in 1885, but at times he expressed his frustration about the illusory nature of scientific and ecological progress. In an 1874 letter to his colleague George B. Emerson, for example, Hough voiced his existential uncertainty: "[I] am collecting a very considerable amount of material and information. But I sometimes stop to ask myself the question—what use can be made of this—or what good can it bring to myself [or] the world?"[63]

Hough participated in the same intellectual community as did Warder and Morton. He endorsed "Arbor Days" and contributed to the same forestry promotion congresses and publications as the other two men.[64] But perhaps because of his preference for introspection and meticulous data collection, Hough never gained a following like Warder's or a movement like Morton's. A polymath among polymaths, Hough dedicated himself to exploring his myriad interests instead of cultivating a group of disciples. In addition to his medical work and his studies of silviculture and climatology, Hough wrote a series of New York county histories and studied the history of American meteorology.[65] He maintained a lifelong interest in lunar and solar halos, sketching thousands of them alongside his daily weather observations. Hough's exhaustive 1857 study "Essay on the Climate of the State of New York" correlated the presence of these phenomena with "the excess of minute spiculae of ice in the upper region of the clouds," thereby affirming the popular view that the halos were "a precursor of rain."[66] Hough also served as a superintendent for the 1855 and 1865 New York State censuses. As he traveled through his home state in the 1850s and 1860s, Hough observed widespread deforestation and grew concerned about potential shortages of timber and fuel.[67]

Hough's experiences in collecting agricultural and climatic data galvanized his advocacy for forest conservation. He would go on to write extensive reports on forestry in the 1870s and early 1880s, but as early as 1863, he expressed his concern about the negative climatic consequences of deforestation in a manuscript on New York meteorology. Citing both popular beliefs and the observations of French scientist Jean-Baptiste Boussingault, Hough wrote that development and clearcutting threatened to destroy forests that equalized his state's climate.[68] One year before the publication of George Perkins Marsh's influential work *Man and Nature* in 1864, Hough reached similar conclusions about land degradation and its climatic effects.[69] The two northeasterners also agreed about what Marsh termed "the uncertainty of our meteorological knowledge." In his 1863 manuscript, Hough confidently described society, trees, water, and air as mutually influential, but he also invited agriculturalists and others to investigate the mysterious workings at the root of any climatic changes. Hough mentioned hilltop groves as especially worthy of future climatological studies.[70]

In terms of influence and public prominence, the New York statistician never matched Marsh or even less famous authors such as Morton. As David Lowenthal points out in his biography of Marsh, Hough greatly admired the "Prophet of Conservation." He "held Marsh the pioneer crusader against excessive felling" and hoped that he would lead the cause of American forestry. The feeling was not mutual, however: Marsh once wrote to the botanist Charles S. Sargent that he did "not expect much from Dr. Hough."[71] Although one of the giants of American conservationism dismissed him as a mediocrity, Hough—an empiricist who grasped the limits of empiricism—possessed a keen understanding of the paradoxical relation between industrializing society, science, and nature. His writings demonstrate that notions of uncertain

ecology predated Marsh's rise to prominence and circulated among a
network of amateur scientists. Hough's work offers a glimpse into the
intricate web of nineteenth-century ecological thought.

Historian Donald Worster identifies two strands of this web, both
originating in the 1860s cultural moment following Darwin's *Origin of
Species* and lasting into the 1890s. Worster characterizes one of the intellectual currents as a Victorian philosophy emphasizing "man's" Dar-
winian dominance over "nature." He describes the second current as
more "biocentric"—as a framework within which humanity played a
small role relative to its surroundings. Worster views Marsh as an ad-
herent of the first intellectual tradition, emphasizing the Vermonter's
belief in the power of society to transform landscapes, often for the
worse.[72] Hough fused the two seemingly contradictory ecological perspectives. On one hand, he wrote that "human agencies sink into insignificance" when compared with vast natural forces such as oceanic
and atmospheric currents. On the other hand, Hough asserted that humans "may have had an appreciable influence" on climatic and ecological systems.[73]

Cleary, uncertainties about scale underlay late nineteenth-century
ecology, climate theory, and forest advocacy. Historian of science
Deborah Coen has demonstrated how, in late nineteenth-century
Austria-Hungary, multiscalar understandings of climate and landscape
dovetailed with an appreciation for ecological interconnections and
complexity.[74] Like Coen's imperial climatographers, Euro-Americans
debated whether local, regional, or hemispheric scales offered the best
framework for understanding and mediating environmental phenomena. Some of Hough's contemporaries described ecological and climatic changes across scales, from local to global. Others emphasized
only local changes. Still others remained vague about the dimensions of
ongoing or potential climatic and environmental transformations.

Writings detailing planetary-scale, human-induced ecological
changes contain an undeniable element of hubris. They appear as the
ultimate expression of the Victorian belief in human dominance over
nature or as a reflection of the halcyon days of the Gilded Age, a period when some Anglo-Americans placed almost no limits on their
visions of ascendancy.[75] Yet many forestry advocates echoed Hough;
they described humanity as humble in scale before the vastness of nature while also upholding human agency as manifest and transformative. They invoked large-scale ecologies not only to laud society's progress, but also to offer dire warnings.

In 1885, for example, Kansas scientist Francis H. Snow described an
inevitable trend beyond the reach of human agency to reverse: "There

can be no doubt that the earth is very gradually approaching the moon's condition, and that some time in the far distant future, how many millions of years hence no man can determine, its atmosphere and surface waters will entirely disappear."[76] Snow was a cautious optimist, however, and he believed "man's influence upon nature" could temporarily counter desiccation and degradation in specific locales. California horticulturist Elwood Cooper sounded a more alarmist note. Cooper's 1876 book *Forest Culture and Eucalyptus Trees* praised Marsh and Hough for their forest advocacy and employed much the same scale as Snow. But in contrast to the Kansan, Cooper characterized the global ecological catastrophe as both anthropogenic and imminent. Although Cooper believed forest culture could mitigate some symptoms of the planet's decline, he described degraded regions as beyond society's ability to repair. The damaged areas, he wrote, would not "become fitted for human use except through great geological changes, or other mysterious influences or agencies of which we have no present knowledge, and over which we have no present control."[77]

Cooper paired uncertain ecology with heady interventionism: he trumpeted the climatic benefits of local and regional-scale eucalyptus planting in California even as he acknowledged the limits of human ecological knowledge and agency. In light of their uncertainty about the role of humanity in relation to "nature," authors such as Cooper and Snow developed a fluid politics of scale. While Hough struggled to devise data collection efforts that ranged across scales, his two contemporaries deftly incorporated multiple scales into their ecologies.

Like the question of spatial scale, the issue of ecological time offers a glimpse into the Gilded Age dialectic between hubris and fatalism, between visions of utopia and fears of catastrophe. Climate theorists and forestry advocates deployed varying temporal scales to many ends, from assuaging anxieties to instilling a sense of urgency. In 1883, the *Garden City (KS) Herald* published a fictional retrospective from one hundred years in the future. By March 24, 1983, the *Herald's* editors imagined, Garden City would be the "Metropolis of South Western Kansas," with 100,000 inhabitants. Its future residents could boast that "through a wise policy of planting and fruit culture the old northerns gave way to zephyrs. Kindly rains visit us, without disastrous floods, or dreaded droughts." The newspaper envisioned a transformed Great Plains ecology, complete with a harmonious climate. The *Herald* may have intended its column to be lighthearted and amusing, but its hundred-year time frame demonstrates a long-term commitment to environmental transformation.[78]

In contrast to the *Herald*, which endorsed the notion that forest culture restored ecoclimatic equilibrium, Adolphus Greely sought to disprove claims that human agency rendered climates either more or less unbalanced. Greely, head of the US Signal Corps, invoked historical memory to question the theory that "floods, tornadoes, and other violent destructive meteorological phenomena" had become "more frequent in this age than in former centuries." Greely's 1890 manuscript pointed toward oral traditions and anecdotal evidence of violent floods and storms from the past, indicating that meteorological conditions had remained consistent.[79] By arguing in favor of continuity, Greely cast doubt on equilibrium-based theories of climatic and ecological change. Greely had access to vast amounts of climate data collected by the Smithsonian and the Signal Corps. But his understanding of ecological and climatic disturbances seems only slightly less speculative than that of a small-town Kansas newspaper, demonstrating that historical interpretations and anecdotes retained their importance amid the increasing quantification of climate science. Five years after Hough's death and after further data collection, ecological and climatic mysteries remained as stubborn as ever.

While Greely and the *Herald* employed centennial time scales, Nebraska scientist Samuel Aughey went even further, traversing millennia and venturing into the realm of astronomy. Aughey's 1877 *Nebraska Farmer* article "The Increasing Need for Forests" claimed that if Mars is "without inhabitants, as some astronomers claim, after having once been peopled, it must be because they were as big fools as a portion of the nations of the old world were, and destroyed all their forests. For nothing tends so much toward desert conditions as the neglect of forest culture." While he lamented the devastation caused by human agency, Aughey also believed that "re-clothing the plain of the west with timber would . . . restore a humidity to a climate that has long since left it."[80] Aughey used the language of restoration because, like many other forestry advocates from Kansas and Nebraska, he believed that at some unknown point in the ancient past the shortgrass prairies of the Great Plains had been a lush, forested landscape. And Great Plains booster-scientists like Aughey were not the only ones who advanced this hypothesis. The 1870s and 1880s saw the continuation of a long-running debate about the origins of the Plains, with some authors arguing that treelessness was "natural" and others contending that the prairies and plains had somehow been reduced to an unnatural barren state.[81] Hough himself repeated the common interpretation that Native Americans had created the "barrens" by setting fires "from time immemorial."[82] In addition to providing another justification for dispossession, the ancient forest theory furnished Euro-Americans with a usable history, an idealized past that could serve as the basis for an ecological restoration project.[83]

But this imagined past was not stable: the ecological and climatic history of the Great Plains puzzled settlers, scientists, and forestry advocates alike.[84] Many disagreed with Hough and Aughey's contention about ancient forests. Henry Inman, the author of an 1878 piece in the *Kansas City Review of Science and Industry*, characterized the region between the Mississippi and the Rockies as "enigmatical and mysterious." After surveying a range of explanations for the Plains' barrenness, Inman concluded that the cause for the "destitution of trees . . . is a purely meteorological one."[85] By depicting the Plains' treelessness as the result of natural meteorological factors, Inman exonerated both Native Americans and anglophone settlers for their purported role in destroying the ancient forests of the Trans-Mississippi West. One year after Inman published his article, J. K. Macomber advanced a similar argument, asking "why the States of Ohio and Indiana should not have been similarly denuded of trees. Those states were also originally inhabited by Indians, who were familiar with the use of fire."[86]

Forestry advocates contested interpretations such as Inman's and Macomber's but never succeeded in building a consensus. Edgar T. Ensign, a US Department of Agriculture official, endorsed restoration theories while moving his time scale closer to the present, blaming denudation both on Native Americans and on the "wasteful and improvident methods" of Euro-American lumbermen.[87] Ensign's 1888 report reflects a diversity of opinion about the origins and causes of the Plains' ecological state. As geologist J. D. Whitney observed in 1876, the controversy over ancient forests and the origin of the Plains "is one possessing a great deal of interest, since there is far from being any unanimity of opinion about the various points which are involved in it."[88] Hough realized that Gilded Age forest advocacy relied on climatic theories rooted in shaky statistical grounds. Climate-based conservationism rested not just on tenuous scientific theses, but on uncertain historical visions as well.

While some forestry advocates studied ecological and climatic changes in historical time, geological time, and planetary scales, others preferred minutiae. Hough analyzed both large-scale phenomena and intricate mechanisms. He strove to grasp the inner workings of ecological relationships, focusing especially on what he described as "reciprocal influences that operate between woodlands and climate."[89] By analyzing local scales and the influence of small groves, Hough believed he could gain a glimpse into myriad ecological relationships. He devised a holistic interpretation, arguing that the creation and de-

Figure 5.2. From *Letter from Dr. Franklin B. Hough* (1885).

struction of forests would influence everything from humidity levels to wind strength to blizzards.[90] To some extent, Hough borrowed from his predecessors and from other conservationists, ecological thinkers, and climate theorists of his era. But he also devised his own idiosyncratic theory of forest influences, one rooted in "filaments." Hough envisioned "filaments" of rain as the key agent shaping the interaction between atmosphere and landscape. He believed rain fell from clouds in thin bands and that groves of trees could help draw these bands toward the earth's surface, mostly by increasing humidity levels but also by promoting upward air currents. Judging from Hough's 1882 explanation of "filaments," it seems he derived his theory at least in part from observation: "In a dry time, we sometimes see filaments of rain descending from a cloud, which dry up and disappear in the warm air before reaching the ground. When such clouds pass over large bodies of woodland, where the temperature is cooler and the air more moist, those filaments extend down and afford a shower of rain, but dry up again as they come to the warm air of the fields beyond."[91] Hough sketched the "filaments" in a letter written shortly before his death (in 1885) to Joseph Nimmo of the Department of Statistics (fig. 5.2).[92]

Other authors besides Hough developed their own theories explaining the interactions between forests and climate. These theories were related—many emphasized the complex ecological interrelationships and the multiple factors shaping climate change—but also divergent. In 1860, for example, John H. Klippart of Ohio discussed "the gaseous exhalations of plants" and their role in shaping air currents "in the middle atmosphere." Like Hough, Klippart believed forests attracted clouds, but he emphasized electrical currents much more than the New Yorker, writing that forests shape "the action of electricity" in the atmosphere and thereby reduce hail and lightning. Klippart also emphasized the climatic influence of the "myriad pores with which the plants are furnished in their every part."[93] Thirty years later, John Hay of Kansas advanced a strikingly similar argument; he connected "stomata" on vegetation to "stupendous" climatic improvement. Hay shared Klippart's belief that plants' pores could absorb and then secrete enough moisture to influence meteorological conditions.[94]

Despite their similarities, Klippart's and Hay's theories testify to the changes in ecological thought that took place between 1860 and 1890. Klippart's piece touches on the role of conifers in containing "miasmic and contagious diseases." Medical geography persisted into the 1890s, but later authors mentioned the medical benefits of trees less often than their predecessors, preferring to speculate about high-modern climate engineering proposals such as Hay's own plan to improve the Great Plains climate with a "vast body of water in artificial lakes, reservoirs and canals."[95] Overall, however, juxtaposing these two pieces highlights continuity, with the two theorists advancing similar ecological visions. Hay's and Klippart's pieces drew from a range of growing disciplines, from botany to medicine to hydrology to physics and the study of electricity. As nineteenth-century Euro-Americans sought answers to questions about climate, forestry, and conservation, they used these disciplines to create a dizzying array of ecological theses. But new disciplines and scientific methodologies did not simply give rise to a stable, usable body of knowledge; they sparked a proliferation of related but disparate theories.

Bernhard Fernow: Rooting Out "Forestry Cranks"

Whereas Hough sometimes seemed paralyzed by the fractured nature of late nineteenth-century knowledge production, Bernhard Fernow did not view ecological or scientific uncertainties as an obstacle.[96] Fernow proved equally adept at parsing chaotic scientific debates and intricate natural systems. In a manuscript written sometime between 1885 and 1900, Fernow surveyed the rugged landscape of climate theory and drew an implicit link between epistemological and "natural" ecologies:

> There has been considerable wild discussion on the influence which forests are supposed to exercise on the climate, waterflow, and other conditions. On the one hand, enthusiastic forestry advocates, who clamor for forest preservation—in itself a misleading term—have claimed extravagantly and unconditionally such influences. On the other hand, men that ought to have had enough scientific training to

know better have as unconditionally and extravagantly denied such influences. The one position is as unphilosophical as the other. Every student of nature, be he only an observer of it in the field or be he only an observer of what has been written by observers, knows that all things are in relation, that therefore we cannot take away anything from complex conditions of nature, without affecting more [or] less all other conditions.[97]

Fernow sought to distinguish himself from heedless boosters of forest culture and shrill obscurantists who opposed conservation. He acknowledged that he did not know how "all things are in relation," but he viewed the "relations" themselves as too crucial for inaction. Ecological interdependence trumped ecological uncertainty. Fernow did not know when sufficient "proof and exact data" about climatic changes would be "brought by scientific method."[98] In the meantime, he believed, it was necessary to promote conservation and implement forest management initiatives. And Fernow proved to be one of the most eloquent and passionate promoters of an interventionist and public approach to forestry. The social and climatic influence of trees, he wrote in a pamphlet, "raise the forest-cover above the position of a mere material resource and establish the right and duty of society, the community, the state, to interfere with its use by private individuals."[99]

Born in Posen, Prussia, in 1851, Fernow was much younger than Warder, Morton, and Hough. His biographer, Andrew Denny Rodgers, described him as "by nature an austere Prussian" who emigrated to the United States by "accident" after his engagement to the American Olivia Reynolds.[100] Fernow worked for the Prussian forest service and resumed his forestry work soon after his arrival in the United States, ascending to the position of head of the Forestry Division in the Department of Agriculture in 1886. His work ethic rivaled Hough's. Over the course of his career, Fernow produced a large collection of pamphlets, reports, and articles. He dedicated himself to institution building, first at the Forestry Division and later during his career as an academic and a forester at Cornell University, Pennsylvania State University, and the University of Toronto. Fernow built on the legacy of Morton, Warder, Hough, and other proponents of forest culture, but he also endeavored to break from his predecessors and create a more "scientific" type of forestry. As a high-ranking bureaucrat in a relatively new bureaucracy and an academic in newly founded forestry departments, Fernow often struggled for legitimacy and funding.[101] Perhaps because of his often-tenuous position, he sought to distinguish himself from hucksters and ascientific charlatans.

Fernow wrote that forest advocacy had been commandeered by "forestry cranks" who "are constantly churning in their mill—a mill in which facts and imagination with an addition of a large quantity of assumption are turned into an indigested and indigestible compound by the name of forestry literature."[102] Fernow trod a fine line. On one hand, he had to disavow "cranks" who threatened his discipline's legitimacy. On the other hand, he constantly addressed the fleeting nature of expert knowledge and of the climatic and ecological theories on which he based his forestry platform. Fernow circumvented this seeming contradiction using a variety of tactics. He derived legitimacy from his probabilistic approach and his mastery of complex, mysterious ecological forces. Fernow also made careful use of language and, in certain instances, avoided categorical statements. By stressing that afforestation rendered precipitation more predictable and less capricious, he could claim that forests shaped climatic conditions while allowing that they might not increase rainfall totals. At times Fernow described controversial claims about forests' climatic influences in legalistic language: "The claims are, that a forest cover tempers like a water surface and to some extent intercepts or reduces the force of the hot and cold winds. With all the consequences of such action, further that it influences, if not in amount, yet in local and temporal distribution, the precipitation of rain and snow, beside exerting minor influences."[103] Circumlocution granted Fernow a measure of plausible deniability. He could invoke uncertain climate theories while still maintaining a safe distance from illegitimate "cranks."

In at least one instance, Fernow credited his predecessors and the Arbor Day movement encouraging a change in popular attitudes about forests. Arbor Day celebrations, he wrote, might seem "inadequate, almost childish," but since the "great mass is moved by emotion" rather than "reason," Arbor Day served to bring about a change in mentality from extraction to conservation.[104] In other instances, however, Fernow could not contain his condescension toward earlier proponents of "forest culture" who guided the forestry movement in the two decades following the Civil War. He commented disdainfully on Hough's amateurism, saying he lacked "technical knowledge." Fernow characterized early "popular writers on forestry" and "friends of forestry reform" as well-intentioned but unworthy of being called scientists—they "generalized without sufficient and relevant premises, and before it was possible for science and systematic observations to furnish groups or sound deductions."[105] In light of his efforts to advance his own approach to forest management, it is not surprising that during the 1890s and 1900s Fernow grew somewhat overzealous in his attempts to distinguish himself from earlier authors.

Some historians of forestry seem to have taken Fernow's assertions at face value.[106] Rodgers, for example, credits Fernow with advancing the cause of "true science" and introducing ecological approaches to American forestry.[107] Yet debates about the legitimacy of forest-based climatic and ecological theory long predated Fernow's career and continued thereafter. Accepting Fernow's distinction between "cranks" and scientists reaffirms other dichotomies, such as those between subjective "forest culture" and objective "forest management," between elite science and popular science, and between the Gilded Age and the Progressive Era.[108]

Authors from the 1860s through the 1880s prefigured Fernow's belief in ecological interdependence, as well as the sense of moralism that permeates his writing. Juxtaposing Fernow's work from the 1890s and 1900s with his predecessors' reveals surprising continuity. Klippart's 1860 piece, for instance, anticipated Fernow's arguments about the social and communal value of forestry. The Ohioan's rhetoric presaged the ethos that would characterize later foresters such as Fernow and Gifford Pinchot—a commitment to the "greater good," an attitude Donald Pisani characterized as "Progressive Moralism."[109] Most strikingly, Klippart employed phrases similar to Fernow's. The two men shared a fondness for digestive metaphors. The Prussian condemned "indigested" forestry literature, while Klippart recommended deploying forest science toward a "political end"—the creation of a "well digested system of forest regulations."[110] Also like Fernow, Klippart described the "indirect" ecological, social, and economic benefits of forests.

"Forest influence on salubrity," Fernow claimed, ranked high among these "indirect benefits." In addition to praising trees for creating climatic conditions unfavorable to "pathological bacteria," Fernow credited groves for the "obstruction of air movements carrying microbes."[111] His use of enduring theories about miasma and medical geography belies claims that he created an entirely new, professional brand of forestry. Many of Fernow's amateur predecessors and contemporaries, including politician and forestry advocate Cassius M. Clay, made similar assertions. Although Fernow's writing signals a growing acceptance of germ theory, notions of forest-induced salubrity stretched back at least to the Early Republic.[112]

According to Fernow, the beneficial "indirect influences" of forests extended beyond microbial mitigation: "A further indirect sanitary influence must not be overlooked in our modern economy of city life. The recuperation of bodily energy and of spirit which an occasional sojourn in the cool, bracing, and inspiriting forest air brings to the weary dweller in the city must not be underestimated in the general health conditions of a people." Fernow's holistic approach was equal parts old and new. His ecological sensibility evokes Gilded Age "forest culture," and his concern with health is rooted in older vernacular science and folklore.[113] But his anxieties about the "modern economy of city life" reflect the rapid industrialization and urbanization that marked the late nineteenth- and early twentieth-century United States.[114]

Fernow's concerns about modern society's alienation from "nature" and his efforts to form hierarchies of knowledge exemplify some of the concerns arising from the advent of American industrial modernity. He was far from the only forestry advocate who sought to distinguish himself from seemingly illegitimate predecessors and contemporaries. In 1885, for example, Massachusetts politician George B. Loring delivered a speech before the American Forestry Congress in Boston invoking the specter of "pseudo-science" and urging the creation of the "best and most thoroughly organized forestry in the world."[115] But efforts to discredit amateur climate theory and ecology dated back at least to the mid-1860s.[116] If Fernow's attempt to delegitimize seemingly subjective ecological paradigms and theories is quintessentially "modern," it also has a long genealogy.

Fernow's career did not mark the advent of a new era of objective, positivist ecoclimatic science and forestry, but perhaps it marked the culmination of uncertain ecology. Fernow encapsulates the contradictions and paradoxes that characterized environmental and climatic thought throughout the Long Gilded Age. Like Hough, he viewed humanity as minute relative to "grand natural phenomena."[117] But he also exalted the power of state-based forestry to improve complex ecological systems. He disavowed amateur science while drawing from a vast epistemological ecology that included little-known figures such as Klippart and famous amateurs such as Clay. Warder and other Fernow predecessors incorporated scientific mysteries and unknowns into their forest advocacy. But Fernow elevated the deployment of uncertainty into an art, crafting policy proposals founded on inscrutable ecologies.

Figures like Fernow bridged nineteenth- and twentieth-century scientific cultures. Despite their links to older environmental traditions, Fernow and his contemporaries advocated for the entrenchment of scientistic modernity. They made increasingly successful claims to intellectual hegemony, helping to solidify the power of state and institutional science while delineating boundaries between experts and nonexperts. At the same time, however, they also articulated an almost postmodern kind of critique, a sense of doubt about modern rationalism that would persist after the dawn of the twentieth century and into the later Progressive Era.

CHAPTER 6

Technocracy and the Mastery of Uncertainty

The Kansas Controversy of 1907

US Weather Bureau director Willis L. Moore probably did not intend to trigger a storm of controversy. After all, it was 1907, long after the panic of 1893 and the droughts of the 1880s and 1890s had dented Great Plains boosterism, long after Dyrenforth and other rainmakers had been exposed as charlatans, long after windmills and scientific "dry farming" had replaced climate improvement as panaceas in the West.[1] Moore likely believed it was safe to publish a government study claiming that human influences had not improved the semiarid climate of western Kansas.

In his 1907 report, the Washington-based bureaucrat sought to avert a resurgence of belief in anthropogenic climate change. The wet years of the mid-1900s had raised the possibility that afforestation or agriculture-based climate theories might regain some of their former prominence. A few Kansans had warned agriculturalists not to fall into the same trap set by optimistic boosters and climate theorists in the 1870s and 1880s. Writing in 1905, T. C. Henry, the former "wheat king," cautioned his fellow Kansans against short climatic memories. Henry had once adhered to the "increasing rain belt" theory, but he had "finally reached the conclusion that if anything could be proven with respect to the rainfall of Kansas, it could be proven that there was a diminishing rainfall."[2] As evinced by his royal epithet, Henry had a flamboyant streak and seemed to relish polemical writing. Moore, by contrast, cultivated an image of himself as a consummate rationalist and empiricist—in other words, as an ideal Progressive Era technocrat.[3] He seemed to view his conclusions—derived from government data—as straightforward and irrefutable. In a series of letters published by Kansas newspapers, Moore summarized the most salient points of his report. He argued that "we should not be misled by present conditions and assume that Nature will always favor us with an abundance of rainfall. I tried to make it plain that we must expect periods of deficient rainfall as well as periods of excessive precipitation. . . . It is a mistake from your people to assume that civilization has changed your climate. It has not." Moore added that "speculators and land boomers are largely responsible for the erroneous information that has been published in regard to certain regions."[4]

Kansans responded to Moore's missives with vitriol. Letters poured into Kansas newspaper offices, with many readers taking Moore's statements as personal affronts.[5] Though both Moore and Henry rejected theories about human-induced climate improvement, the Weather Bureau director inspired a far more acrimonious reaction than the Kansan, largely because of his status as a government bureaucrat and expert. An unattributed letter to the *Topeka Daily Herald* invoked the material progress of Kansas settlers to disprove Moore's scientific report. The author of the piece took aim not just at Moore, but at all East Coast scientists: "They may reason till doomsday that climates cannot change. They will not convince the Kansas pioneer. Theories will not stand in the way of incontrovertible fact. The whole history of the west says that climates do change." Using folkloric language to underscore his disdain for elite eastern scientists, the unnamed writer likened technocrats to donkeys: "Until the weather experts at Washington can find some better argument than bald assertion and unfounded theory with which to back up their warnings they had better quiet down or expect to be consigned to a place with their fellows in the corral of 'Rocky Mountain canaries.'"[6] The *Kansas City Journal* published a similar letter from R. James Abernathy. Echoing the anonymous author from the *Herald*, Abernathy juxtaposed Moore's dire statements with the "thousands of happy and prosperous farmers living west of the ninety-ninth principal meridian." Abernathy denounced Moore's statements as an "unprovoked official defamation of one of the great states of the Union" and claimed that Kansas had "ground for protest against officious meddling by the head of a department of the government." The climate theories Abernathy espoused showed a remarkable similarity to the climate-improvement theses prevalent in the 1870s and 1880s. Like the writings of Wilber, Aughey, and Elliott, Abernathy's letter maintained that the "growth of trees and cultivation of the soil" had "transformed" the climate of western Kansas.[7]

Moore's detractors harked back to earlier Gilded Age arguments and narratives in other ways as well. Their letters demonstrate that the same

cultural, social, and economic tensions at the heart of the climate debates of the late nineteenth century still inflamed the Plains in the early
twentieth century. Mrs. A. I. Soper, a Kansas resident who had home-
steaded in Beaver County, Oklahoma, accused the Weather Bureau
chief of serving the interests of large ranchers. "Apply the spurs to the
hide of Willis L. Moore," she wrote, "and it wouldn't surprise me any
if it would be developed that the cow men had furnished the data for
the statement made by Mr. Moore." Soper recounted how ranchers had
tried to dissuade her from farming in the marginal areas around the one
hundredth meridian. Like many other letter writers, she relied on the
evidence of her experience in the "fine farming country" of the shortgrass prairie to push back against Moore's claims.[8] Soper's 1907 letter is
strongly reminiscent of 1870s antiranching tracts by Uriah Bruner and
other proponents of Great Plains agriculture.

M. E. Nichols's invective against Moore also employed older tropes
and theories, albeit with a distinctive turn-of-the century flavor. Nichols began his letter by excoriating the government scientist using anti-
intellectual stereotypes, portraying Moore as an out-of-touch expert who
draws a "fat salary" to "sit in his office" and read weather maps. Nichols
then pointed to the recent dispossession of land from Native American
reservations as proof and cause of climate improvement in Kansas:

> In '93 the Cherokee strip was opened for settlement and three-
> fourths of it put into cultivation, and later in '01 the Kiowa and Co-
> manche Indian country was opened to settlers that have since put
> most of it into cultivation, and that all of these things have caused the
> drilling of wells and building of tanks and pools which are kept con-
> stantly filled with water creating thereby an abundance of moisture
> in the atmosphere that would in itself go a long way toward checking
> the hot winds that formerly wrought such havoc in Western Kansas.[9]

Back in 1889, Adolphus Greely wrote that the "confining of Indians to
reservations" would likely bring about a climatic improvement in the
Great Plains. By 1907, "confining" Native Americans no longer sufficed.
Climate improvement and the expropriation of Native American land
continued to function as mutually justifying rationales in the eyes of
Indian-hating expansionists. But after the turn of the century, other
technologies of settlement—including windmills, wells, and retaining
ponds—began to supplement plowing and tree planting in the climatic
and genocidal visions of some Great Plains writers.[10] And the expropriation of government-granted "Indian country" replaced the creation
of reservations as a motivation for climate improvement.

Despite their own use of arguments dating back to the first decades after the Civil War, Nichols and other critics of Moore also accused the Washington expert of having outdated views regarding the Plains. Nichols wrote that Moore "has shown himself by his late comments" to be "far behind the times" by maintaining that Kansas's climate remained dry and unchanged. Similarly, Philip Campbell, the US congressman representing the Third Kansas District, ridiculed Moore for his seeming adherence to antiquated notions about Kansas. "He probably still indulges in the delusion that we have long whiskered Populists," claimed Campbell. According to the *Kansas City Star*, the congressman "roasted" the Weather Bureau in an extensive public rebuke of Moore, even proposing to abolish the organization.[11] Campbell accused Moore of stereotyping Kansans as recalcitrant Populists while employing the same East-West animus and pro-agriculture rhetoric that had been the hallmark of the People's Party.[12] Although by 1907 the Populists had long since been absorbed into the Democratic Party, class tensions clearly continued to animate environmental and climatic discourse.

Moore, for his part, must have been taken aback by the scale and animosity of the backlash: in early March 1907, two months after the controversy began, he issued an official statement to clarify the intent of his initial report. Moore wrote that he had said "nothing derogatory" about Kansas. The newspapers pounced. The *Topeka Daily Herald* of March 9, 1907, included the headlines "Never Said It," "Moore Denies Attacking Western Kansas," and "Says He Was Misquoted." But on the issue of climate change, Moore stuck to his original line. He repeated his claim that "the rainfall has neither increased nor diminished by amounts worthy of consideration" and issued a "solemn warning against the wholesale settlement of that country"—meaning western Kansas and eastern Colorado.[13] Though the attacks against Moore eventually subsided, most Kansas newspapers and letter writers remained unconvinced by the bureaucrat's entreaties.

A 1912 piece in the *Kansas City Star* indicates that resistance to Moore's climatic skepticism remained staunch. Titled "Weather Fakes: Popular Fallacies about Meteorological Phenomena," the article initially seemed to cast doubt on theories of human-induced climate improvement by equating them with the basest of meteorological myths, such as the "rain of frogs." The *Star* paraphrased Moore as saying that "the rain belt of Kansas has not moved one mile further to the westward than it was a million years ago." Surprisingly, however, the piece then began to challenge Moore's claims, arguing that his statements "have been widely questioned" and that "when he made them some time ago he stirred up a hornet's nest of opposition among the residents

of the dry sections of the West, and many adherents of the theory that
the rain belt is moving West have used the records of the United States
weather bureau itself to prove their assertions. These records in many
instances appear to prove that the rainbelt actually is moving West."
The 1912 article from the *Star* illustrates how even Moore's most powerful evidence—his government rainfall statistics—could be turned
against him.[14]

Uncertain Technocracy

The Willis Moore Kansas dispute has been largely overlooked by climate
scholars and historians of the American West. Since it underscores the
persistence of climate change discourse, the 1907 controversy does not
corroborate the often-repeated assertion that Euro-Americans abandoned climate improvement theories after the 1890s.[15] To some extent
the clash between Moore and his critics in Kansas mirrors conflicts between hunters and bureaucrats in one of the well-chronicled metanar-
ratives of Progressive Era conservation—the struggle between "locals"
and increasingly powerful rationalist technocrats.[16] But a closer examination of the 1907 episode reveals a more complex story.

At least one farmer, for example, stood up for Moore. A. E. Comes
of Sedan, Kansas, wrote a letter to the *Kansas City Journal* arguing that
"Mrs. Soper and Mr. Abernathy" were being "ridiculous" in their attacks
on Moore and should "sit down . . . and not speak again until . . . spo-
ken to." Comes emphasized his background to demonstrate that not all
agriculturalists were united against Moore: "I am a farmer, was always a
farmer, and from 1878 to 1888 I farmed in Western Kansas." People with
"selfish motives," Comes wrote, had misled him about the agricultural
prospects of the western portion of the state, and after ten "lost" years
on the shortgrass prairie, he had retreated to somewhat more favorable
climes (Sedan is located east of Wichita and well to the east of the one
hundredth meridian). Invoking the power of fact and the bogeyman of
the speculator, Comes concluded that the Weather Bureau chief had
been unjustly attacked: "Mr. Moore has simply set out existing facts
which should be considered carefully by honest homeseekers. Boomers and wildcat speculators to the contrary notwithstanding."[17]

As for Moore himself, he was neither a quintessential positivist technocrat nor the hardened opponent of climate change theory that his opponents alleged. During his tenure as director from 1895 to 1913, Moore
sought to systematize and rationalize the Weather Bureau's mapping
and reporting networks. Throughout the Kansas climate controversy,
Moore juxtaposed his agency's objectivity against Kansans' purport-

edly unreliable local knowledge. But in some instances, Moore also allowed for the coexistence of weather lore and quantitative meteorology and climatology.[18] Historian Jamie Pietruska has shown that Moore's career was characterized by a gradual acceptance of uncertainty as a necessary component of scientific practice. After spending the first decade of his time as Weather Bureau director railing against speculative and uncertain scientific practices, Moore grew increasingly tolerant of long-range weather forecasts and probabilistic methodologies.[19]

Moore's extensive "Report on the Influence of Forests and on Floods," published three years after the Kansas controversy, reveals that he derived legitimacy not just from rationalism and empiricism, but also from his mastery of indeterminacy. Moore began the report by acknowledging his own changes of opinion on the issue of forests' hydrological and climatological influences. Although he hoped his 1910 report would put to rest the perennial question of forest-induced climate change, he still reserved "the right to change or still further modify my views if the presentation of new facts and figures render[s] such a course logical, and do not consider that I shall stultify myself in so doing." When addressing Kansans during the 1907 controversy, Moore portrayed his findings as permanent truths. In his 1910 report, by contrast, he emphasized the fleeting and temporary nature of any scientific conclusion.[20]

Moore erred on the side of caution, voicing his strong skepticism of regional-scale climate change theories but refusing to rule out local-scale human influences on weather patterns. His report took on an irreverent and perhaps playful tone when discussing scale, even indulging in a seeming digression on agriculture: "The covering of tobacco plants with thin cheese cloth results in establishing a local climate which will continue so long as the cloth remains in position."[21] Similarly, Moore went on, "the erection of a tent, of a barn, of a dwelling house, of a village, or the growth of a great city . . . influence[s] the local climate in proportion to the area that is covered." Any categorical disavowal of climate change theses remained impossible, Moore believed, because climatic and meteorological phenomena needed to be analyzed across a multitude of scales. Moore seemed comfortable with his conclusions even as he allowed that "a *statistical* solution of this problem" remained illusory.[22] If Moore—the head of a major bureaucracy who often presented himself as a consummate technocrat—infused his empirical studies with irreverence and doubt, perhaps uncertainty was present at the very fountainhead of Progressive Era positivism and empiricism.

Moore's writings during and after the 1907 climate controversy highlight the obstacles faced by early twentieth-century technocrats. Progressive experts sought to make truth claims and effect policy using an often tenuous knowledge base. They operated in a cultural climate rife with contestation, in an arena where myriad voices such as those of Abernathy, Soper, and Nichols could erode and cast doubt on any consensus. In the early twentieth century, the question of human influences on climate remained as vexed and unsettled as ever. Yet experts still needed to address it in their efforts to justify studies and policies related to forestry, hydrology, agriculture, and resource management more broadly. Instead of relying solely on positivism and notions of efficiency, a number of technocrats made recourse to uncertainty when faced with climatic quandaries. Some, like Moore, gestured toward temporary statistical uncertainties that could be overcome with time and further research. Other authors acknowledged deeper ontological uncertainties and unknowns.

As scholars like Ian Tyrrell have demonstrated, Progressive Era conservation began as a reaction against the uncertainties created by urbanization, industrialization, the misuse of resources, and the perceived closing of "virgin lands" in the West.[23] Perhaps the relation between conservationist technocracy and uncertainty ran even deeper. Progressive Era experts embraced mystery and produced unstable knowledge, proving that technocratic paradigms both reacted against uncertainty and depended on it.[24] In *Rebirth of a Nation*, T. J. Jackson Lears characterizes Progressivism as "torn by tensions—populism vs. expertise, producerism vs. consumerism, statutory vs. administrative regulation." Lears identifies another tension in turn-of-the-century culture: that between the inexorable advance of rationalist scientific management and the emergence of modernism and antipositivism, two intertwined intellectual currents that "seemed to be exploding all the old metaphysical certainties."[25] By employing both positivism and uncertain science, conservationists attempted to reconcile the opposing currents Lears describes. Their efforts show that technocratic modernity originated not just from rationalism and positivism, but from embracing uncertainty as well.[26]

At the Intersection of Hydrology and Climatology

Rationalist technocrats derived much of their legitimacy from their ability to manage air, water, and land. For Moore and his colleagues in other government bureaucracies such as the Forest Service, the Geological Survey, and the Corps of Engineers, the emerging discipline of hydrology took on paramount importance. In the decades spanning the turn of the century, state and federal bureaucrats dedicated themselves to the study and management of water resources, often serving the interests of private entities such as large-scale irrigationist and agricultural enterprises.[27] Myriad reports and studies detailed the hydrologic benefits of forests, justifying the continued efforts of foresters and bureaucrats. Only efficient and scientific forest management, the logic went, could ensure the continued availability of water. "The most important of all functions of the [forest] reserves is their yield of water," proclaimed the "Friends of the Forests" at a meeting in Denver in 1901.[28] Bureaucrats sought to apply their hydrologic expertise to both the East and the West. They focused their efforts on landscapes they viewed as degraded by deforestation and erosion. Although recent studies have shown the conservation-preservation schism to be something of a false dichotomy, other fractures divided the burgeoning progressive technocracy, and hydrology offered some measure of consensus.[29] Both "wild" forest reserves and carefully controlled plantations would regulate river flow, while both small farmers and large corporate interests would benefit from reliable stream discharge and the mitigation of droughts and floods.

In addition to promising consensus, hydrology offered the allure of certainty. Water engineering schemes epitomized progressives' attempt to "simplify the world" and rise above the uncertainty and unpredictability of complex landscapes.[30] Some turn-of-the-century technocrats attempted to shift environmental discourse away from unsettled climatological questions and toward hydrological studies. Whether writing a scientific treatise, a speech, or articles for the popular press, several experts and officials employed a similar rhetorical strategy: juxtaposing a dubious statement about forests' effect on climate with an authoritative claim about the benefits of efficient hydrological management. In 1901, for example, government hydrologist Frederick H. Newell told an audience of Nebraska horticulturists that "while man could not change the wind or the rainfall much could be done to conserve the rainfall which nature gives us."[31] Focusing on quixotic climatic theories, Newell implied, would only distract from the concrete subject of watershed management.

Newell's strategy likely appealed to his fellow hydrologists. But most other bureaucrats, whether they worked for the Forest Service, the Geological Survey, or the Corps of Engineers, could not resist the allure of climate change theses. They could not flout an environmental tradition that dated back to G. P. Marsh's *Man and Nature* (1864), if not earlier.[32] Rather than attempting to separate hydrologic questions from climatic ones, many technocrats chose to work at the intersection of climate science and water management. Fusing water management science with climate theory allowed technocrats to retain older forms of environmental holism while also appealing to popular beliefs about climate change. Operating at the nexus of hydrology and climatology afforded progressive technocrats a measure of flexibility before the public. When seeking to encourage public scientific forest management in the forested East, for example, they could warn about desiccation and the intertwined hydrological and climatic effects of deforestation. When focusing on the treeless portions of the West, on the other hand, they could emphasize the virtuous cycle of climate improvement and water abundance created by afforestation. Often, progressive technocrats sought to simplify the natural world. But just as often they were content to harness the mysterious and interlocked forces of air and water.

The hot summer of 1901 illustrates why conservationists could not let go of tenuous climatic theses. The sweltering days of July 1901 saw an outpouring of articles and reports discussing the possible role of society in exacerbating the heat wave that had struck much of the United States. The *Eureka Californian* asserted that the "connection between deforestation and drouths is direct and unquestioned. So much has been established beyond the peradventure of a doubt."[33] The Eureka paper did not clarify exactly how deforestation caused drought. Was it by directly influencing atmospheric conditions? By diminishing the soil's ability to retain rainfall? Or by influencing both hydrologic and climatic conditions? Despite, or perhaps because of, its certain tone, the *Californian* felt no need to add specifics about the mechanisms at the root of environmental and climatic changes. Eastern papers, meanwhile, also entered the discussion. The *Boston Evening Transcript* challenged the *Springfield (MA) Republican*'s skepticism about the role of denudation in causing the recent spell of hot weather. After allowing that questions of causality remained uncertain and "but slightly understood," the editors of the *Transcript* argued that it was "safe to say" that forest cover played a role in the heat wave and associated drought. "Tests by the Weather Bureau," the Boston paper added, "have shown that the mean temperature is lower at the stations in the wooded districts than at those on the open plains."[34] Newspapers clearly did not let scientific uncertainty prevent them from making claims about society's influence on the environment.

Perhaps sensing that the 1901 heat wave offered a singular opportunity for advancing their programs, two of the leading figures of Progressive Era conservation decided to add their voices to the discussion. Secretary of Agriculture James Wilson prefaced his statement with a major disclaimer: "I do not know exactly what direct influence the presence or absence of forests has on the climate, but I know this—that the dev- astation of the ranges is resulting in the growth of the arid lands of the West. . . . And as the arid lands increase it is reasonable to believe the hot winds will increase."[35] Whereas Wilson focused on the West, Gif- ford Pinchot, head of the Bureau of Forestry and the leading light of conservationist technocracy, set his sights on the East and Midwest, de- scribing how vast areas had been cleared in the name of agriculture. "It is impossible to say," Pinchot stated, "just how much effect this clearing process has had upon the temperature and climate" in formerly wooded areas, "but we believe if there were more forests and timber land ex- treme heat and cold would be avoided."[36] By using the word "impos- sible," Pinchot signaled that the forces shaping climate and hydrology were not just temporarily uncertain, but perhaps unknowable as well.

Neither Pinchot nor Wilson made explicit claims about forests' in- fluence on rainfall totals or humidity levels in the wake of the 1901 heat wave. Both depicted aridity, water availability, forestry, and climate as interconnected without elucidating a clear causal mechanism or expla- nation. Even if forest management could not directly increase rainfall levels or lower temperature, perhaps it could trigger a virtuous hydro- logic and climatic cycle. Wilson and Pinchot also resisted the urge to enter into a dispute with newspaper editors who criticized scientists for using the drought to galvanize support for their policies. Neither seemed to take the bait, for example, when, a few weeks after they is- sued their statements, the *Manchester (NH) American* issued this re- buke of technocratic expertise: "Because the scientists say so, and because it looks reasonable, we are bound to believe that drouths and freshets are due to the destruction of the forests and resulting exposure of the land to the rays of the sun, but when we try to bolster this the- ory with facts from experience we run into those that do not confirm it."[37] Unlike Moore, who would attempt to fight fictions with facts in 1907, Pinchot and Wilson seemed to grasp that many facts had a lim- ited cultural life span, even in a time of growing faith in empiricism and rationalist management. At least in this instance, the two eminent technocrats embraced the uncertain nexus of hydrology and meteorol- ogy in their public discourse. They chose to display their comfort with unstable knowledge instead of making authoritative claims and risking becoming embroiled in a controversy as Moore would a few years later.

Figures such as Pinchot and Wilson exerted a strong influence on lower-ranking technocrats in the Progressive bureaucracy. But their views never acquired the power of gospel, since even middling figures felt the need to reopen the question of forests' influence on climate and hydrology when writing their reports. The very structure of progressive technocracy seemed to perpetuate uncertainty. In their need to publish ever more scientific papers and local case studies, prominent and rank-and-file bureaucrats gave rise to a dizzying panoply of theories, many of them admittedly tentative.[38] Adding to the confusion, neither Pinchot nor any of his rivals and colleagues ever assigned a single author or department to the task of expounding the government's ultimate stance on human-induced climate modification. The closest approximation to a comprehensive report dated back to 1893: a lengthy study titled *Forest Influences* published by the USDA Forest Service. The 1893 report included contributions by some of the most well-respected scientists in the employ of the US government, including Bernhard E. Fernow of the Division of Forestry as well as Cleveland Abbe, George E. Curtis, and Mark W. Harrington—three of the government's leading experts on climatology and meteorology. Despite their wide-ranging knowledge, the authors of *Forest Influences* could not resolve the debate about trees' role in shaping climatic conditions. Fernow summarized the report's findings by way of an agricultural metaphor: "The crop of incontrovertible facts is still scanty, and further cultivation will be necessary to gather a fuller harvest and then to set clear the many complicated connections connected with this inquiry." Much like Newell in 1901, in 1893, Fernow deferred to more certain hydrologic evidence, claiming "much more confidence" when addressing the "effect which forest cover exerts upon the disposal of water supplies."[39] Other technocrats seemed to interpret Fernow and his coauthors' uncertainty as a call for further research, and their near-constant flow of reports over the next several decades would not lead to any definitive conclusions.

Surprisingly, however, the conservationists' piecemeal, kaleidoscopic approach proved successful in diffusing the notion that forests could ease hydrologic and climatic crises. Even as the debate about forests, hydrology, and climatology took on all the characteristics of an arcane academic debate, popular publications echoed the technocrats' rhetoric. An 1894 piece in the magazine *Youth's Companion* seemed to refer directly to the massive 1893 *Forest Influences* report. The didactic children's weekly offered a pithy overview of the findings of Fernow and his colleagues: "It appears that, while clearing off the wooded areas affects but slightly the total annual rainfall, forests are invaluable conservators of it. They act as reservoirs of moisture, which they distribute to adjacent fields in rivulets of dew."[40] The technocrats' indecision and uncertainty did not prevent their message about hydrologic and climatic stewardship from proliferating broadly in popular culture. As Kevin Armitage has argued, in the decades around the turn of the century, "the scientific worldview entered the daily lives of people as never before." But the advance of "technological rationality" was not the only factor in bringing about the change Armitage described.[41] In addition to empirical facts, agents of the state also disseminated confusion, revealing the dialectic between uncertainty and rationalism at the heart of the making of technocratic modernity.

The Weeks Act of 1911

The notion that scientific certainty governed all Progressive Era conservationism dates back to Samuel Hays's 1959 *Conservation and the Gospel of Efficiency*. Although many have criticized Hays's work, his argument about scientific rationalism has retained much of its power.[42] Some historians have continued to assert that all conservationists shared a "core set of assumptions," among them the belief that science could "unlock" nature's secrets.[43]

But nature's secrets were just too numerous, at least according to George W. Rafter. Rafter wrote several works on hydrology, including a 1903 US Geological Survey report, "Relation of Rainfall to Run-off." After admitting that "no general formula is likely to be found expressing accurately" the relation between climate and hydrology, he claimed that the only constant to be found was the infinite variability of nature: "As a general proposition we may say that every stream is a law unto itself."[44] Rafter also addressed the role of forest cover in controlling rainfall and runoff. "The evidence on this point is conflicting," Rafter wrote, adding that "the answer to this question must be regarded as very uncertain." Tellingly, he deferred to an 1877 report by Franklin B. Hough and claimed that the twenty-six-year-old report "may be accepted as expressing the fact at the present day." Rafter concurred with Hough's cautious hypothesis about "reciprocal influences that operate between woodlands and climate."[45] Rafter's USGS report underscores the persistence of both forest-climate theory and scientific uncertainty into the early twentieth century. Yet Rafter and his colleagues also faced different challenges relative to Hough and other predecessors from the 1870s and 1880s. Whereas Hough dedicated himself to studying climate and forestry with an eye toward forest advocacy, bureaucrats like Rafter endeavored to transform uncertain theories into concrete hydrologic and environmental policy. Climate change theory had been deployed before to advance legislative acts, especially by boosters and expansionists who pushed for the Timber Culture Act of 1873, but the expansive Progressive state granted technocrats like Rafter greater proximity to the policy-making assembly line.

The technocrats' imperative to transform tenuous hypotheses into practice took on new urgency in the years leading up to the passage of the Weeks Act of 1911. Previously, all national forests had been designated in swaths of public land in the West. The Weeks Act sparked contention because it aimed to maintain reliable watersheds and river flows by allowing the government to purchase forest lands along navigable watercourses throughout the United States.[46] The Weeks Act proved so controversial that it created fractures even within the federal bureaucracy. Some high-ranking members of the Corps of Engineers, a federal agency under the umbrella of the US Army, believed the act would grant too much power to the Forest Service and jeopardize the future of the Corps.[47] Since hydrology served as the Forest Service's justification for purchasing land, debates leading up to the passage of the bill focused on the purported benefits of forests. Opponents of the act voiced their skepticism about trees' influence on both climatic and hydrologic conditions. Proponents of the bill, on the other hand, claimed that forests regulated stream flow, preventing destructive floods and damaging droughts. Both factions made strategic use of doubts and unknowns. Perhaps because of the high stakes, or perhaps because of the sheer weight of the bureaucracy involved, the technocrats deliberating the Weeks Act theorized uncertainty to a greater extent than had many of their predecessors, creating a taxonomy of uncertainty.

Hiram M. Chittenden led the charge against the Weeks Act. Chittenden had worked for the Corps of Engineers since 1884 and by 1909 had ascended to the rank of lieutenant colonel. Over the course of his career, he supervised a number of projects, including constructing tourist roads in Yellowstone Park and building a major canal in Seattle.[48] Chittenden has been described by historian Matthew Klingle as an "astute politician" and a figure who "carried himself like a paragon of Progressive Era virtue, a man who claimed to weigh the facts in acting for the greater good." Chittenden believed engineers needed to "discipline" and reform disobedient rivers. A firm supporter of American exceptionalism and the perpetual renewal of Manifest Destiny, he also wrote several books, most prominently a history of the fur trade.[49] Chittenden's 1909 paper "Forests and Reservoirs in Their Relation to Stream Flow" questioned the link between forest cover and flood control. By challenging the justification for the Weeks Act, Chittenden adopted a seemingly contradictory stance: he tried to make it clear that his "sympathies are wholly on the side of the present movement for the conservation of our national resources" while also challenging one of the core tenets of progressive environmental thinking. The lieutenant colonel took issue with the moralistic assumptions at the heart of conservation: "If this fact of deforestation has brought with it in greater degree than of old the calamities of high and low waters, then, indeed, we are in an unfortunate case. But it has not done so. Nature has decreed no such penalty for the subjugation of the wilderness, and on the whole these natural visitations are less frequent and less extensive than they were before the white man cut away the forests." Whereas most conservationists sought to redress the perceived environmental sins of earlier generations, Chittenden denied that any sins had been committed.[50]

Surprisingly, however, Chittenden did not issue a categorical denial of the hydrologic and climatic benefits of forests. Instead, he attempted to fit specific claims about forests' influences into an epistemological schema. Some theses, he wrote, persisted because "little effort is made to consider whether there may be some other and more satisfactory explanation," and others because the "necessary evidence" was "so hard to get." Still others endured because "the elements of the problem are so many and conflicting" or because they could be neither validated nor disproved within the reasonable future: "To establish the definite falsity of these propositions is an extremely difficult task." Not content with merely pointing out that his opponents' theories were too tenuous to justify passing legislation, Chittenden elaborated on the different categories of uncertainty faced by his foes.[51]

Supporters of the Weeks Act faced a greater challenge than Chittenden did. Even if they proved as adept at theorizing uncertainty as their rival from the Corps of Engineers, proponents of the legislation still needed to convince politicians and the public that their tentative hypotheses warranted the public purchase of private lands. A 1913 report offers perhaps the clearest glimpse into the methods and strategies of the technocrats who favored the Weeks Act. The bill was passed in early 1911, but its final version contained a stipulation that the Geological Survey conduct more definitive studies before authorizing the purchase of any lands. The 1913 study, which—like seemingly all Progressive Era reports—had a predictable and prosaic name, *The Relation of Forests to Stream Flow*, played a key role in retroactively justifying the Weeks Act and allowing for its implementation.[52]

The study's authors included Benton MacKaye of the Forest Service and Mashall O. Leighton, the chief hydrographer of the Geological Survey. Leighton, MacKaye, and their collaborators carried out a case study in the White Mountains of New Hampshire intended to prove a correlation between forest cover and river discharge. Leighton and MacKaye employed divergent strategies when presenting their conclusions. Leighton's portion of the report, focusing on "hydrometric studies," attempted to isolate unproven theories about climate from the report's hydrologic evidence: "Forests may or may not induce copious precipitation. The fact is of no importance to this report. Let it be emphasized that this paper is confined to a study of the relation of forest cover to run-off." Echoing Rafter's aphorism that no two streams were alike, Leighton stressed that even his hydrologic findings could not be extrapolated to understand other regions' watersheds.[53]

MacKaye, on the other hand, sought to convince his readers that uncertain hypotheses played an intrinsic role in scientific practice. He resorted to probabilistic language, hoping skeptics would be swayed by an appeal to risk and chance: "The probabilities are that the closer measure of forest influence is based on the hypothesis including the factor of flow efficiency rather than on the hypothesis excluding that factor."[54] Despite their sometimes tortured language and circuitous explanations, MacKaye's and Leighton's reports helped legitimize the Weeks Act and create the White Mountain National Forest. Their success belies the notion that rigid scientific positivism pervaded Progressive Era culture. To some extent, uncertainty helped technocrats like Leighton and MacKaye gain traction with both politicians and the public. If science served as the guiding spirit of American conservationism, then progressive science must have been capacious—or at least expansive enough to include all the permutations of uncertainty described by Chittenden, Leighton, and MacKaye.[55]

Climatic Morality Revisited

The debate surrounding the 1911 Weeks Act highlights the inadequacy of the label "technocrat." Though I have employed the term throughout this chapter, figures such as MacKaye and Chittenden cannot be pigeonholed as cogs in the progressive bureaucratic machine. A self-described supporter of conservation, Chittenden doubted one of the most basic precepts of the conservation movement: the notion that humans served as a "disruptive force in nature, almost always upsetting its harmonies."[56] In Chittenden's view, Euro-Americans' moral imperative to transform landscapes had not been imperiled by capitalist expansion. He flouted the guiding ethos of progressivism by arguing that unchecked economic development and environmental improvement could go hand in hand. MacKaye transcended the boundaries imposed by the growing Progressive bureaucracy in a different manner. For MacKaye, the conservationist environmental ethic needed to be paired with a commitment to the politics of social and economic equality. He attended Harvard University's forestry school and gained the specialized knowledge essential to technocratic governance while also maintaining his commitment to radical politics. Like his involvement with the "Hell Raisers," a group of radical reformers, MacKaye's vision for the egalitarian "community settlement of wild lands" shows that his wide-ranging interests spread far beyond the study of climate and hydrology.[57]

MacKaye's and Chittenden's political leanings diverged starkly: Chittenden exalted development while MacKaye censured individualistic expansionism. But both infused their scientific reports with moralism. According to Armitage, many Progressive Era conservationists reacted against the culture of instrumental rationality produced by modern science. Because of its "lack of moral content," empirical science could not convey the moral urgency many reformers desired. Armitage argues that many conservationists struggled to couple "the separate worlds of moral and scientific certainty."[58] Some, as he has shown, sought a reprieve from rationalist bureaucracy in communities such as the experiential Nature Study movement. Uncertain science offered another way of circumventing the quandaries presented by positivism. By forgoing certainty in his hydrologic writings, MacKaye honed a scientific vision that befitted his moral uncertainty about the broader economic and political system of Progressive Era America. Chittenden, by contrast, used epistemological mystery to reaffirm the moral validity of expansionism and development.

Writers from beyond the federal bureaucracy also invoked climate theory while expounding on the merits and problems of American society. In the wake of the 1901 heat waves and droughts, for example, the *Hubbard (IA) Monitor* sounded a note similar to MacKaye's. The newspaper cited unnamed scientists' claim that "an oak tree eight inches in diameter will exhale into the atmosphere during the season two hundred tons of water." "If that estimate is anywhere near true," the *Monitor* explained, "it is easy to see why the summer might get hotter each year." The paper's editors acknowledged that self-serving experts may have exaggerated their claims, but they still used the exhalation hypothesis to level a broadside against powerful industrialists and their political allies. According to the *Monitor*, tree planting would benefit all members of society by regulating climate patterns. Yet politicians continued to "wantonly [waste] money on river and harbor" improvements that "benefit only a few individuals who have mills and factories beside the streams."[59] The *Monitor* combined populist antielitism with progressive utilitarianism, a philosophy that would soon be encapsulated by Pinchot's 1905 credo, "the greatest good, for the greatest number, for the longest run."[60]

While Pinchot stood as the paragon of conservation, figures like MacKaye were atypical and idiosyncratic but still very much within the circle of elites at the heart of progressivism. Writings such as Pinchot's and MacKaye's tell an incomplete story. Indeed, bureaucratic histories fail to convey the full breadth of Progressive Era conservation. Climatic and environmental debates waged in newspapers, pamphlets, and other publications paint a picture that is fuller, if still incomplete. Euro-Americans of various stripes prescribed tree planting and climate improvement as panaceas for myriad environmental, economic, cultural, and political ills. In their struggle to bring tenuous climatic theories into debates about capitalism and political economy, authors mixed old tropes with newer notions emerging from turn-of-the-century culture.[61]

In 1907 Willis Moore encountered resistance from across Kansas society, from people who viewed climate improvement as a reflection of their hard work and progress. Others, in both East and West, warned about climatic deterioration, following the tradition of older Gilded Age authors who depicted droughts, floods, violent storms, and unpredictable weather patterns as retribution for society's infractions. The *Jacksonville (FL) Times Union*, for instance, depicted recent "devastating droughts" and "extraordinary" rainfalls as a "silent but solemn admonition of nature which we shall do wisely not to disregard." Deforestation and conservation, the paper argued, exerted only a negligible influence on the "vast continental air movement." But on a local and regional scale, trees functioned as an "equalizer of evaporation, of cloud-formation and of rainfall."[62] Two years later, in 1901, the *Nebraska City (NE) Conservative* published a letter personifying nature in much the same way. J. R. Lowell, the author of the letter, cautioned that "Nature may not instantly rebuke, but she never forgives a breach of her laws." Sounding like a progressive technocrat, Lowell wrote "that the influence of trees upon climate and rainfall gives to the planting of trees, and to the protection of them where nature has already planted them, a national importance." It is improbable that Pinchot monitored the small-town Nebraska press, but Lowell's admonitions would have warmed his heart: "Our wicked wastefulness and contempt for the teaching of science in this matter will most surely be avenged on our descendants."[63] Viewed together, the writings from Nebraska City and Jacksonville highlight two Progressive Era dialectics: one between elite science and the popular press and another between deep-seated concerns about environmental degradation and continued faith in capitalist expansion.

"Science" appears in Lowell's writing as a fount of authoritative knowledge. By appealing to "science," he may or may not have been referring to reports and studies published by technocrats like MacKaye, Rafter, and others. Yet Lowell's letter indicates that not all residents of the Great Plains resisted institutional science as much as Moore's opponents would in 1907. Often, the claims of government experts found their way into the popular press, albeit in reconstituted form. Studies about forests' climatic and hydrologic benefits acquired new meanings in the Great Plains, where they could be used to support renewed efforts to settle marginal agricultural areas.

Change and Continuity

While the Great Plains continued to function as a test site for environmental and climatic expansionism, old notions of agricultural development took on new forms during the age of technocracy. The panic of 1893 cast a long shadow. A financial crisis that sent shock waves throughout the US economy, the panic purged some climate improvement narratives of their boosterish tendencies. In the immediate aftermath of the crash, figures such as Martin Mohler attempted to reconcile beliefs about climate change with stark new realities. Mohler's 1893 paper, "A New Departure in Agriculture," illustrates how climate theories continued to circulate, though as somewhat chastened versions of their predecessors. Mohler had both the financial panic and recent droughts in mind when he acknowledged that the most ebullient climate theories had proved "a delusion and a snare" for settlers.[64] But the former Pennsylvanian and active member of the Kansas Board of Agriculture still clung to the belief that human agency could improve climatic conditions: "While it is true that there has been in twenty-five years no perceptible increase in the rainfall of central and western Kansas, yet it is also true that as cultivated areas have been extended and forests have been multiplied, climatic conditions have been improved for agricultural purposes." The more moderate version of climate theory Mohler espoused was congruent with the careful, tentative claims of government scientists. By arguing that human influences had improved general climatic conditions without increasing rainfall amounts, Mohler echoed forestry experts like Fernow.[65]

Mohler sought the broadest possible audience for his theories; he read his paper before the Agricultural Congress at the 1893 World's Columbian Exposition in Chicago. Speaking against the backdrop of the fair's merrymaking, he sounded a somber note, explaining that the "collapse of the boom" had been six years in the making and had recently been exacerbated by falling prices. Mohler urged his listeners to take some comfort from the boom-and-bust "Ferris wheel"—an apt metaphor considering that the world's first Ferris wheel had been unveiled at the fair. He believed the panic and the concurrent droughts had made agriculturalists more cautious; busts had rendered their enterprise more suited to long-term growth and prosperity.[66]

Despite Mohler's continued faith in expansion, Frederick Jackson Turner, who also attended the World's Fair, issued his iconic remarks about the closing of the frontier. Turner's "frontier thesis" has generated well-founded and much-needed critiques from generations of historians.[67] Back in the 1890s and early 1900s, however, proclamations like Turner's spurred a different reaction. Some writers took great pains to prove that the 1880s droughts, the panic of 1893, and other developments had not sounded the death knell of westward expansion. Whereas Mohler reaffirmed the belief that climate improvement would facilitate settlement in arid regions, other authors, such as Judge S. Emery, looked to irrigation as a means of perpetuating agrarianism. In an 1897 article "Our Arid Lands" in the popular publication the *Arena*, Emery wrote that the clouds left by the crash had finally lifted. Immigration, "held in check by the period of depression," had resumed. And, Emery claimed, "once again the people are talking of going West. The beginnings of a new Western movement are plainly discernible." Emery betrayed his status as an irrigation booster, arguing that only irrigation would remove "the utter uncertainty of crops" that had haunted farmers in the High Plains and other portions of the arid and semiarid West. Mention of climate improvement was conspicuously absent from Emery's article. Settlers, he argued, should not accept the West and its "scanty natural rainfall," because irrigation would "redeem" arid lands.[68]

Alongside irrigation, windmills began to supplant afforestation-based climate improvement in expansionist rhetoric. Windmills granted Great Plains farmers the ability to pump seemingly limitless supplies of water from underground aquifers. Once above ground, the water could be circulated and conserved using systematic irrigation systems.[69] Although ranchers had used windmills for decades, their popularity expanded in the 1890s as newspapers proclaimed the advent of "Irrigation's Boom."[70] Irrigation proponents downplayed the prospect of climate improvement to make their technologies seem all the more miraculous. In 1894 the *Topeka Daily Capital* profiled farmers who "look below instead of above for crop moisture." The *Capital* claimed that irrigation could bring agriculture to previously "impossible" areas.[71]

Sensing an opportunity to put their theories into practice, government experts joined the push to create a more reliable, scientific brand of agriculture. The Department of Agriculture dedicated itself to studying and promoting "dry farming," an agricultural doctrine rooted in scientific soil culture and the use of special tillage techniques to conserve soil moisture. As the economic historians Gary Libecap and Zeynep Hansen have explained, dry farming dovetailed with the progressive technocracy's "belief in the practical use of science to advance human welfare." Thanks to the support of government agricultural experiment stations and "Dry Farming Congresses," scientific dry farming proliferated across the West after the turn of the century.[72] In the wake of droughts and economic panic, windmills, irrigation, and dry farming offered a previously unimaginable level of certainty. The dawn of a technological, rationalist era seemed at hand. Proponents of westward expansion, it appeared, would no longer have to rely on tenuous and folkloric climate theories.

But confident pronouncements about the advent of a new agricultural age contained a hint of desperation. Kansas and Nebraska would never experience another demographic boom quite like the railroad-fueled frenzy of the 1870s and early to mid-1880s.[73] As the focus of farming settlement shifted toward the northern Great Plains of Montana and the Dakotas during the 1900s and 1910s, homesteaders continued to try their luck on the prairies.[74] And Great Plains farming remained a risky undertaking. Despite the advantages conferred by windmills, aquifer-based irrigation, and scientific soil conservation, the possibility of farm failure still loomed, especially in marginal areas averaging less than twenty inches of annual rainfall. Questions about the long-term tenability of small-scale farming persisted, with some critics viewing the trend toward larger landholdings as a possible solution.[75] Notions of human-induced climate improvement still held a great appeal in the uncertain context of the early 1900s Great Plains. Indeed, irrigation boosterism and the new emphasis on scientific dry farming did not so much replace climate modification as supplement it.

Some irrigation proponents ridiculed climate modification initiatives. After moving away from Kansas to get "into the irrigation ditch business in Colorado," T. C. Henry, the former "wheat king" and proponent of Kansas agriculture, mocked Kansans for continuing to believe in climate improvement. In a 1901 letter to the *Wichita Eagle*, Henry snidely remarked that Kansans would receive more rainfall only if they managed to "build some mountains" like Colorado's.[76] Newell, the federal hydrologist and irrigation proponent, mirrored Henry's message as well as his tone. Newell's 1903 report trumpeting the potential of irrigation in the West parodied those who believed in the westward movement of the "rain belt." Describing eastern Colorado, western Kansas, and western Nebraska, Newell wrote, "This wonderfully attractive and in many ways rich country may be called the famine belt. In it many attempts have been made, in vain, to secure permanent settlement, and thousands of industrious and hard-working settlers have been forced to leave by starvation."[77] In contrast to Henry and Newell, however, some climate improvement proponents devised theories compatible with the spirit of the turn-of-the-century irrigation boom. They argued that ponds and other water storage reservoirs used in irrigation would ameliorate the climate of arid regions. "Pond theory," as some called it, represented the synthesis of long-standing, uncertain, and unproved climate theories with the instrumental, technoscientific ethos of the irrigation era.[78]

Pond theory derived from older notions about the possible use of artificial bodies of water in climate modification schemes. During the 1870s and 1880s, James Humphrey, Richard Stretch, and others set the stage for later developments by theorizing that artificial inland seas and pond construction could add significant moisture to the atmosphere. Even John Wesley Powell, who was hesitant to endorse any climate improvement beliefs, speculated that irrigation canals and other artificial bodies of water might start a virtuous hydrologic and atmospheric cycle in the West. Writing in 1888, Powell claimed that "irrigation will increase the humidity of the climate, and increase protection from fires to the non-irrigated lands; and, as the lands gain more and more water from the heavens by rains, they will need less and less water from canals and reservoirs."[79] With the increasing focus on irrigation after about 1890, many Great Plains writers trumpeted claims like Powell's. An 1890 letter to the *Topeka Daily Capital* called for Congress to be "beseiged [sic]" until it provided support for afforestation and pond construction. The author of the letter, credited only as "Zac" of Dodge City, Kansas, invoked "practical knowledge" in explaining that the "moist air above ponds is heavier than its surroundings." The "heavy air," Zac believed, would help "insulate areas from hot winds" and create conditions more conducive to agriculture.[80] Writings by Zac and by M. E. Nichols, one of the anti-Moore letter writers from 1907, show substantial popular support for pond theory.

Following Powell's lead, some government officials and "men of science" endorsed pond theory as well. In an 1893 speech, for example, John R. Sage articulated a climate amelioration program that featured both ponds and afforestation: "With great timber belts judiciously located, and by the construction of artificial ponds and lakes, we may in time be well-nigh exempt from damage by hot winds and drouths, and we may even ward off a large measure of danger from tornadoes."[81] By claiming that human agency could mitigate extreme weather events, Sage tapped into a long-standing belief dating back at least to the 1870s.

Sage added a new ingredient to the mix by including pond theory. As director of the Iowa Weather and Crop Service, Sage usurped Gustavus Hinrichs's position as the leading figure in Iowa climatology and meteorology. According to Sage's side of the story, Hinrichs had run his Iowa State Weather Service on "an independent line" and had thus squandered the opportunity to "secure the benefits of reciprocity and co-operation with the National Weather Service."[82] By entering the orbit of the federal bureaucracy, Sage hoped to increase the "scope and efficiency" of the Iowa weather service. He envisioned a new organization capable of producing an unending stream of usable empirical knowledge. Sage assumed control of the Weather and Crop Service in 1890, about the time when many Iowans had become embroiled in a debate about the possible climatic consequences of tile drainage.

Some alleged that tile drainage, a technique used to drain boggy areas and reclaim fields from swamps and marshes, had caused an increase in aridity across Iowa.[83] Sage put "no stock in that sort of gloomy prognostication" and reaffirmed expansionist tropes about humanity's power to improve environments.[84] If anything, he claimed, tile drainage served "to *increase* rather than lessen the capacity of the soil to retain moisture." In Sage's view, drainage actually added moisture to the atmosphere and increased rainfall levels. The head of the Weather and Crop Service thus found himself in a somewhat puzzling position: he had to contend that artificial ponds added moisture to the atmosphere while also arguing that natural ponds and bogs needed to be drained because they removed humidity from the air. Adding to the chaos, Sage admitted that neither pond construction nor tile drainage nor any other factor had thus far caused an observable change in Iowa's climate. "The records," he wrote, "show no evidence whatever of a change of climate in this western region." On the one hand, Sage trumpeted the power of artificial ponds and forests to control extreme weather conditions. On the other, he put "little stock in any theory that implies a very large measure of human power over the elements."[85] Judging from Sage's writings, pond theory offered no more certainty and clarity than afforestation- or agriculture-based climate hypotheses. Sage hoped to bring rationalism to Plains climate science by toeing the line set by his superiors at the US Weather Bureau. But like Hinrichs, his predecessor from the 1870s and 1880s, Sage soon developed a facility with conflicting data and confounding climate theses. He set out to rebuild Iowa climate science on a more efficient, positivist model only to find that modern technocracy required an aptitude for contradiction and uncertainty.

Shelterbelts and the Long History of Climate Theory

Perhaps because of its compatibility with irrigationist rhetoric of the 1890s and early 1900s, pond theory flourished around the turn of the century. Yet nobody attempted to implement pond theory on a massive, continental scale. Although irrigationists constructed many ponds and reservoirs, they did not participate in a concerted national climate improvement program. The history of shelterbelt planning followed a similar arc until the 1930s, when New Deal technocrats put their theories about windbreaks into action.

Like pond theory, the notion of using belts of trees to modify wind currents dated back to the nineteenth century. In 1873, William Hammond Hall's report on Golden Gate Park described how groves of trees would serve to regulate both wind speed and precipitation patterns.[86] Sage himself mentioned the climatic potential of "judiciously located" timber "belts" in his 1893 speech. Progressive Era Shelterbelt proponents emphasized trees' ability to mitigate damaging winds more than their power to transform precipitation patterns. They traded the utopian appeal of grandiose schemes for inducing precipitation for the more immediate and tangible appeal of shelterbelts. Whereas Sage and Hall trumpeted trees' power to avert tornadoes and regulate rainfall, shelterbelt proponents such as Isaac Cline, E. F. Stephens, and H. C. Price (writing in 1894, 1897, 1902, respectively) described the microclimates within shelterbelts' wind shadow. Rows of carefully planted trees would reduce evaporation and lessen wind erosion. Shelterbelts, these authors argued, could grant farmers oases free from the vicissitudes of the harsh Great Plains climate.[87]

Historians Robert Gardner and Joel Orth have chronicled budding attempts to construct shelterbelts and other artificial forests in the Great Plains during the 1890s and early 1900s. Their studies reveal the myriad challenges and complexities encountered by practitioners of afforestation. Gardner, for example, describes federal efforts to construct artificial forests in Nebraska as a Progressive endeavor characterized by a focus on "centralized control over production, including labor, location, process, and decision-making." Gardner also explains that foresters such as Charles E. Bessey came to view artificial forests as "an evolving environment that was of their own making yet beyond their full control."[88] After the turn-of-the-century attempts Gardner describes, shelterbelt initiatives proceeded only by fits and starts until the Dust Bowl, when the US Forest Service and the Department of Agriculture initiated the massive Shelterbelt Project. Inaugurated in 1934, the project attempted to construct a 1,200-mile band of forests extending from

Texas to North Dakota. As Orth explains, the Shelterbelt Project aimed to hold back the "drought, dust, and despair of the Dust Bowl."[89]

Although the ecological and social cataclysm of the Dust Bowl imparted urgency to the Shelterbelt Project, the planning and implementation of the project proved just as contentious as earlier attempts to carry out conservationist policies. The controversies surrounding the New Deal shelterbelt mirrored those that took place in the lead-up to the Weeks Act of 1911. Again, grandiose visions for environmental and climatic conservation encountered resistance from the public and from within the federal bureaucracy. In 1933, chief forester Robert Stuart proposed a modest plan that focused on shelterbelt planting "along highway and section lines throughout the Plains and prairie regions of the Midwest."[90] Raphael Zon, another high-ranking government forester, advanced a far more ambitious project. As the leader of what Orth termed the "pro-climate faction," Zon claimed that the climatic benefits of forests justified creating a more extensive shelterbelt. His opponents feared that Zon would resuscitate antiquated ideas about climate improvement and jeopardize the Forest Service's hard-won legitimacy and scientific credibility. But President Franklin D. Roosevelt seemed taken by the Zon faction and favored a more grandiose plan.[91]

Zon's writings on conservation, climate, and forestry display a strong consistency with the work of earlier technocrats and climate-improvement proponents. An émigré from the Russian Empire and, in the words of historian David Moon, a "strong-willed character" who "relished argument," Zon worked closely with Pinchot and Fernow.[92] Over the course of the 1920s and 1930s, Zon published a series of studies, some intended for the broader public and some for scientific audiences. In terms of epistemology and scientific philosophy, Zon maintained that "divergence of opinion" and persistent scientific uncertainties should not preclude the implementation of conservationist policy. In terms of climate theory, Zon reified long-standing beliefs about forests' power to transform climate. Though he recognized Willis Moore as a "high meteorological authority," he took issue with Moore's claim that precipitation originated largely from ocean currents.[93] Zon espoused a bold version of forest-climate theory: he even claimed that during the summer, when westerly winds gave way to southeasterly winds, forests in the southeastern United States determined the precipitation patterns for the "central states"—"the granary of the United States." Zon felt comfortable publishing these claims even though "direct proof of this climatic influence quantitatively expressed is still lacking."[94]

Zon, one of the architects of a major New Deal initiative, accepted uncertainty as an unavoidable component of modern science and tech-

nocratic environmental governance. Only some of Zon's contemporaries and predecessors shared his approach, but his writings indicate
that climate change and uncertainty continued to shape conservation
discourse during the height of the New Deal era. New Deal conservation exerted a profound influence on postwar America, spreading belief
in the welfare state and shaping popular views on nature.[95] Conservation initiatives born from uncertain technocracy continued to shape
American society long after the Dust Bowl era.

But why did climate change belief persist so long? And how did uncertain climatology and forestry survive all the dramatic transformations
that shook American culture and society from 1870 to 1930? To some
extent, climate theory remained intertwined with doubt and mystery
because, as the *Iowa Farmer and Breeder* asserted in 1892, the vastness
and complexity of climate systems rendered any "practical demonstra-
tion" of theory impossible.[96] In a broader sense, perhaps, some characteristics of American modernity reinforced uncertain epistemologies.
The burgeoning but chaotic bureaucratic state, the proliferation of
newspapers and scientific societies and journals, the economic panics
that sent shock waves through society, the moral and social quandaries
confronted by MacKaye and others, the ecological pressures exerted by
westward expansion, industrialization, and urbanization—all these developments unsettled efforts to create certain, stable knowledge. In the
uncertain climes of modern America, figures such as Zon, Chittenden,
and Rafter thrived. They flourished because they proved adept at creating, navigating, and coping with mysteries and unknowns.

The Meanings of Uncertainty

Much like the work of Gilded Age and Progressive Era climate theorists, this project remains incomplete and leaves much untrodden ground. Mysteries abound.

Perhaps this book's greatest lacuna is its lack of transnational scope. The history of American environmental and climatic thinking cannot be separated from the history of global empire.[1] Climatic hopes and anxieties shaped Euro-Americans' continental and global visions. In 1885, for example, M. O. Baldwin wrote "The Panama Canal and Its Possible Influences upon the Ocean Currents and upon Climate," published in the *Kansas City Review of Science and Industry*. An unabashed supporter of the then-hypothetical Panama Canal, Baldwin claimed that the artificial waterway would modify "the great thermal currents from the Pacific" and change the course of the Gulf Stream. Puncturing the isthmus, he argued, would allow wider circulation of mild air currents and render harsh subarctic regions habitable. It is unclear to what extent Great Plains climate modification debates inspired Baldwin's grandiose claims about Panama. Yet his writings highlight the link between climate change theories and broader imperial aspirations.[2]

Scientific debates about climate change reached beyond borders. Hough, Fernow, and other forestry advocates belonged to an international community of climate theorists who participated in a lively exchange of ideas.[3] Especially during the 1860s and 1870s, the writings of Germans Ernst Ebermayer and Heinrich Wilhelm Dove exerted a strong influence on American authors. As early as 1857, Hough noted that Dove had "beautifully" characterized society's relation to the natural world as a constant struggle between the "principle of destruction" and that of "preservation."[4] Ebermayer conducted a series of experiments aiming to measure the atmospheric influence of forests.

Published in 1873, Ebermayer's report on his findings garnered attention from several American authors, including Hough. Ebermayer's work remained relevant into the twentieth century, earning citations from meteorologist Cleveland Abbe in 1905 and forester Raphael Zon in 1927.[5] In addition to these two Germans, French and Italian forest-climate theorists influenced American thinkers, as did British colonial foresters working in South Asia.[6] And climatic ideas flowed in both directions. American proponents of climate improvement invoked the authority of A. I. Woiekoff (Voeikov), a prominent scientist from St. Petersburg University. Woiekoff seemed to return the favor by endorsing the rain-formation theory devised by George Curtis, the federal meteorologist who accompanied Dyrenforth on his expedition to Midland.[7] Many climate change theorists sought legitimacy from their links to Old World scientists, but it remains unclear whether invoking European science had the desired effect. The international network of climate change theorists merits closer examination: To what extent did European contemporaries stabilize or unsettle American scientific and environmental discourse?

I also remain uncertain about the nature of the relationship between Gilded Age environmental thinkers and their precursors. Territorial expansion, environmental change, and economic dislocation created climatic concerns in the seventeenth, eighteenth, and early nineteenth centuries.[8] How did the scientific and cultural doubts from earlier eras influence Gilded Age discourse? Were Gilded Age climatic uncertainties the result of the particular conditions of late nineteenth-century American industrialization and urbanization? Or did all colonial settlement projects give rise to intertwined visions of climatic utopia and catastrophe?[9] Is it possible to identify a pattern of ebb and flow between certainty and uncertainty in the longer history of American scientific cultures?

The unsettled scientific and cultural climate of the late nineteenth century underscores how questions of irreducible uncertainty originated long before the advent of "reflexive modernity," postmodernism, and the "post-truth" era. The capitalist transformations of the nineteenth century exposed Americans to what Jonathan Levy described as "radical uncertainty."[10] Gilded Age and Progressive Era figures such as Hinrichs, Roberts, and Hough understood that modern efforts to mitigate risk, mystery, and complexity gave rise to new uncertainties.

During America's first Gilded Age, making recourse to scientific certainty rarely resolved complex environmental and epistemological disputes. In our own historical moment, seemingly intractable problems such as climate change have exposed the limits of positivism and straightforward facts. The recent history of global warming belies the notion that scientific facts alone will lead to appropriate political action.[11] Perhaps it might be useful to rediscover more uncertain traditions of environmental and scientific thinking. My aim is not to exalt late nineteenth-century climate theorists as environmental and scientific heroes. Morton, Wilber, and others sometimes endorsed reprehensible aspects of Euro-American expansionism. But they seemed to grasp a concept recently articulated by Mike Hulme: the argument that "far from being able to eliminate uncertainty, science—especially climate change science—is most useful to society when it finds good ways of recognizing, managing, and communicating uncertainty."[12] Some nineteenth-century figures also subscribed to the notion that treating science as an apolitical and discrete body of knowledge could limit its efficacy.[13] They grasped the power of acknowledging and incorporating uncertainty into climate treatises that fused science and politics.

As climate "skeptics" have shown, the use of uncertainty can be strategic, cynical, and nefarious. Uncertainty, however, has many meanings and implications. It can also connote a humble recognition of the persistence of scientific and historical mystery.

ACKNOWLEDGMENTS

In the years I have spent reading and writing about various forms of uncertainty, I have become more and more certain that I am deeply indebted to a large number of people.

Many librarians and archivists helped me find and access historical documents. At the American Geographical Society Library, Jovanka Ristic and Marcy Bidney helped me track down countless nineteenth-century climate and weather maps. Cheryl Oaks, Eben Lehman, and Lauren Bissonette provided invaluable assistance in helping me sort through the rich holdings of the Forest History Society. I am especially grateful that they provided scans of the Dyrenforth experiment images. At the American Philosophical Society, Roy Goodman shared an extensive bibliography of climate history sources he compiled along with historian James Rodger Fleming. In Albany, Paul Mercer of the New York State Library helped me navigate the Franklin B. Hough Papers. Lauren Gray of the Kansas Historical Society provided some crucial newspaper sources.

I am also grateful to the generous institutions that granted funding. At Cornell University, the History Department, American Studies Program, and Institute for Social Sciences enabled me to conduct research and write. I am deeply appreciative of the American Meteorological Society and its funding of humanities research. The American Meteorological Society Graduate Fellowship in the History of Science allowed me to carry out archival research from 2014 to 2015. The American Geographical Society, the Forest History Society, and the New York State Library supported me with travel grants that allowed me to spend time with their extensive collections.

I am tremendously lucky to have Aaron Sachs as a mentor. If not for his generosity, brilliance, and support, this book would not exist.

Aaron's tremendous knowledge of nineteenth-century culture improved the depth and breadth of my research and writing. In both my teaching and my writing, I constantly ask myself what Aaron would do in my situation. And of course, I would be irresponsible if I did not conclude this paragraph by saying, "Go Sox!"

Sara Pritchard and Ray Craib gave me support and sage guidance on innumerable occasions. Sara's comments on early chapters broadened the scope of my research, while her advice on later chapters expanded my historiographic engagement. Ray's influence on my thinking is especially evident in chapter 4.

I am grateful to Daegan Miller, Aaron Sachs, Molly Reed, Amy Kohout, and many others for creating and sustaining HAW! (Historians Are Writers!), an incredibly supportive community. Daegan also gave me crucial advice on a very early draft of my introduction. Amy's enthusiastic support and deep knowledge of the US West expanded the range of my research and writing. Molly's work on utopian communities influenced my work in surprising and always positive ways.

I first began to develop the idea for this book in a seminar paper for Maria Cristina Garcia's graduate course on twentieth century US history. In addition to Professor Garcia, many other Cornell University professors and graduate students read and commented on early chapter drafts, including Brian Rutledge, Tim Sorg, Max McComb, Laura Martin, Fritz Bartel, Jacob Krell, Jack Chia, Alana Staiti, Robert Travers, Ellen Abrams, Matt Dallos, Christine Croxall, Molly Geidel, Benedetta Carnaghi, Jackie Reynoso, Alberto Milian, Mark Deets, Ryan Edwards, Ronald Kline, Durba Ghosh, Josi Ward, Mattias Fibiger, Ernesto Bassi, Joshua Savala, Paul Nadasdy, Nick Bujalski, Kyle Harvey, Kelly King-O'Brien, Sean Cosgrove, Alex Vo, Amanda Bosworth, Nate Boling, Nick Myers, Ryan Purcell, Edward Baptist, Kevin Bloomfield, and Enzo Traverso. Barb Donnell's support helped me through every phase of graduate school.

Back in my undergraduate days at Middlebury College, some extraordinary professors shaped my thinking so deeply that their presence is still felt in this book. Febe Armanios and Ian Barrow made me want to be a historian. Anne Knowles's energetic and inspiring classroom presence made me want to be a teacher. Anne's historical geography courses were a revelation that completely changed my understanding of what historical inquiry might entail.

Andrew Amstutz has had a huge influence on this book. He read and commented on every chapter, always offering incisive and generous suggestions. From Ferrara to Ithaca to many places in between, Andrew has been an amazing friend.

My friends and colleagues at NYU Shanghai and Duke Kunshan University provided support through the final stages of the writing. Amy Goldman inspired me to finally finish this book by organizing a writing group along with Anna Greenspan and Christina Jenq. Amy's remarkable generosity also helped me and my family feel at home in this huge city. I am so lucky to have met Adam Yaghi, a brilliant, kind, and hilarious friend. Adam's support helped keep me sane through many months of teaching and writing. In Shanghai, I have also benefited from the support and collegiality of Jun Yang, Jennifer Tomscha, Mark Brantner, Amy Becker, Maria Montoya, Genevieve Leone, Dan Keane, Ezra Claverie, Lin Chen, David Perry, Sophia Rizzolo, and Cian Dinan. In Kunshan, many people welcomed and helped me during these strange and uncertain times. I am especially grateful to Kolleen Guy, James Miller, Jesse Olsavsky, Yifei Qi, and Kaiyue Sun. Countless students invigorated my thinking in many ways. I am lucky to have worked with so many dedicated students whose ideas prompted me to reexamine my assumptions about history.

I am grateful to Tim Mennel for seeing potential in this project back when it was in its early, inchoate stages. Tim's patient and careful guidance helped me shape the manuscript into more cohesive form. Several anonymous reviewers also offered incisive and constructive comments. At the University of Chicago Press, Susannah Engstrom, Alice Bennett, and Beth Ina helped guide the book through the final stages.

Outside academia, Liam Brenner and Tom Hall supported me with their friendship and humor. As I wrote most of these chapters, I was lucky to live within driving distance of Owais Gilani and Talha Ali—two extremely generous friends. Many others offered a welcome respite from reading and writing: Evan Dalton, Bob Nicholson, Gibo Pezzotti, Alessandro Botticchio, Lizz Huntley, Matt Walker, Mark Mentele, Tim Dellett, and Richard Fink. Joe Atencio was a true history aficionado. His passing has left a huge void in many people's lives. Rosa Xia helped care for my daughter during many months of manuscript revisions. I am thankful for her hard work and her patience.

My family has been a constant source of love, support, laughter, and inspiration. I couldn't have finished this project without them. My grandfather, Joseph Innamorati, passed away as I started working on this project. It was a gift to be able to spend so much time with him. Mary and Doug Bellemare, Rosemary Bellemare, and Madeline and Jared VanSpeybroeck have been exceedingly supportive, especially during an unexpected five-month stay in Jacksonville, Florida. Hillary Giacomelli and Matt Longman: You guys are the best. I've lost track of how many times you and your crazy cat Franklin hosted me at your apartment in Brooklyn. I would never have made it to any archives and conferences without your generosity. Thanks for all the good times! It is not easy to live so far from my father, Savio Giacomelli. But it is always a pleasure when we can get together and set a skin track through the snowy fields and larch forests of Caalér. In Breno, Mina Mossoni and my grandmother Lidia Bersani always lend a helping hand. Grazie di cuore! More than anyone else, my mother, Jean Innamorati, has shaped my intellectual growth. By instilling in me a love of reading and learning, my mother inspired me to try to become a historian. For that, and for so much more, I am deeply grateful.

Sarah Bellemare uprooted her life multiple times, moving across states and oceans, to help me pursue my quixotic obsession with the past. She has made a lot of other sacrifices on my behalf, such as enduring my never-ending barrage of writing-related questions. Sarah's love and sharp intellect kept me from "careening off through the dim mists of cloudland." And she can make me laugh like no one else. Finally, I thank little Louisa—the ultimate source of uncertainty! Louisa's joyful presence has made everything so much better.

Portions of the introduction appeared previously in Joseph Giacomelli, "The Meaning of Uncertainty: Debating Climate Change in the Gilded-Age United States," *Environment and History* 24, no. 2 (May 2018): 237–64. An earlier version of chapter 3 appeared in Joseph Giacomelli, "Unsettling Gilded-Age Science: Vernacular Climatology and Meteorology in the 'Middle Border,'" *History of Meteorology* 8 (2017): 15–34.

NOTES

Introduction

1. William Hammond Hall, "Influence of Parks and Pleasure Grounds," *Biennial Report of the Engineer of the Golden Gate Park, for Term Ending Nov 30th, 1873*; reprinted in *Overland Monthly* 11, no. 6 (December 1873): 535 (pagination is from the *Overland Monthly* version).

2. For an example of this more straightforward improvement ideology, see David Nye, *America as Second Creation: Technology and Narratives of New Beginnings* (Cambridge, MA: MIT Press, 2003), 9, 110, 205. According to Nye, Americans envisioned the natural world as "awaiting fulfillment" through human action.

3. Hall, "Influence of Parks," 527.

4. Hall, "Influence of Parks," 527. For more on Golden Gate Park see Richard Walker, *The Country in the City: The Greening of the San Francisco Bay Area* (Seattle: University of Washington Press, 2007), 59–60.

5. For examples of older transnational debates about climate, see Richard Grove, *Green Imperialism: Colonial Expansion, Tropical Island Edens and the Origins of Environmentalism* (Cambridge: Cambridge University Press, 1995), and Lydia Barnett, "The Theology of Climate Change: Sin as Agency in the Enlightenment's Anthropocene," *Environmental History* 20, no. 2 (April 2015): 217–37. Nineteenth-century settlement and empire building projects in arid or semiarid portions of Russia, North Africa, Australia, and elsewhere also spurred debates about climate change, desiccation, deforestation, and climate improvement. See David Moon, *The Plow That Broke the Steppes: Agriculture and Environment on Russia's Grasslands, 1700–1914* (Oxford: Oxford University Press, 2013); Diana K. Davis, *Resurrecting the Granary of Rome: Environmental History and French Colonial Expansion in North Africa* (Athens: Georgia University Press, 2007); Stephen Legg, "Debating the Climatological Role of Forests in Colonial Victoria and South Australia," in *Climate, Science, and Colonization: Histories from Australia and New Zealand*, ed. James Beattie, Emily O'Gorman, and Matthew Henry (New York: Palgrave Macmillan, 2014); Meredith McKittrick, "Talking about the Weather: Settler Vernaculars and Climate Anxieties in Early Twentieth-Century South Africa," *Environmental History* 23, no. 1 (January 2018): 3–27.

6. Nye discusses the notion of "latent potential" in *America as Second Creation*.

7. For a discussion of climate debates in the colonial and Early Republic eras, see

Anya Zilberstein, *A Temperate Empire: Making Climate Change in Early America* (New York: Oxford University Press, 2017); James Rodger Fleming, *Historical Perspectives on Climate Change* (Oxford: Oxford University Press, 1998); Fleming, *Meteorology in America, 1800–1870* (Baltimore: Johns Hopkins University Press, 1990). In the context of the US West, notions of human-induced climate improvement had circulated at least as far back as the 1840s. See John F. Freeman, *High Plains Horticulture: A History* (Boulder: University of Colorado Press, 2008), 52.

8. See, for example, Frank R. Kimball, "The Climatology of the United States," *Bulletin of the Essex Institute* 18 (January-March 1886): 27–34. Kimball described increasing climatic volatility in California and the northeastern US as evidence of a global phenomenon.

9. For a source discussing how journals "teemed" with climate theories, see Thomas Meehan, "Forests the Result and Not the Cause of Climate," *Prairie Farmer* 44, no. 47 (November 22, 1873).

10. Henry Nash Smith, "Rain Follows the Plow: The Notion of Increased Rainfall for the Great Plains, 1844–1880," *Huntington Library Quarterly* 10 (February 1947): 192.

11. See David Emmons, "Theories of Increased Rainfall and the Timber Culture Act of 1873," *Forest History* 15 (October 1971); R. Kutzleb, "Can Forests Bring Rain to the Plains?" *Forest History* 15, no. 3 (October 1971): 6–14; David Emmons, *Garden in the Grasslands: Boomer Literature of the Central Great Plains* (Lincoln: University of Nebraska Press, 1971); Walter Kollmorgen and Johanna Kollmorgen, "Landscape Meteorology in the Plains Area," *Annals of the Association of American Geographers* 63, no. 4 (December 1973): 424–41; Kenneth Thompson, "Forests and Climate Change in America: Some Early Views," *Climatic Change* 3 (1980): 47–64. For a popular contemporary piece, see Matt Simon, "Fantastically Wrong: American Greed and the Harebrained Theory of 'Rain Follows the Plow,'" *Wired*, https://www.wired.com/2014/06/fantastically-wrong-rain-follows-the-plow/ (accessed July 5, 2020).

12. William Ferrel, "Note on the Influence of Forests upon Rainfall," *American Meteorological Journal: A Monthly Review of Meteorology and Allied Branches of Study* 5, no. 1 (February 1889). Ferrel argues that the afforestation of the Mississippi Valley would increase rainfall levels "as far as the Atlantic Ocean." For a discussion of human-induced changes in rainfall distribution, see Bernhard Fernow Papers, Cornell University, box 2, folder 23, "The Forest as a Condition of Culture," undated paper (possibly an address or presentation). The exact date in unclear, but folder dates attribute it to 1885–88 or 1892 (most likely the earlier dates), 12.

13. The railroad-sponsored booster Richard Smith Elliott acknowledged that "human action has apparently modified conditions, and mostly for the worse, in many parts of the earth's surface." See Richard Smith Elliott, *Notes Taken in Sixty Years* (St. Louis, MO: R. P. Studley, 1883), 300.

14. George Perkins Marsh, *Man and Nature*, ed. David Lowenthal (1864; repr., Seattle: University of Washington Press, 2003), 20–28; Henry Allen Hazen, "Variation of Rainfall West of the Mississippi River," *Signal Service Notes* 7, US War Department (Washington, DC: Office of the Chief Signal Officer, 1883), 3.

15. For one of the most certain voices in the climate debate, see Charles Dana Wilber, *The Great Valleys and Prairies of Nebraska and the Northwest* (Omaha, NE: Daily Republican Printing, 1881), 70.

16. Deborah Coen has analyzed the question of scale in the context of climate history. See Deborah Coen, *Climate in Motion: Science, Empire, and the Problem of Scale* (Chi-

cago: University of Chicago Press, 2018); Coen, "Imperial Climatographies from Tyrol to Turkestan," *Osiris* 26, no. 1 (2011): 45–65. See also Stephen Kern, *The Culture of Time and Space, 1880–1918* (Cambridge, MA: Harvard University Press, 1983), 8; Matthias Heymann, "The Climate Change Dilemma: Big Science, the Globalizing of Climate and the Loss of the Human Scale," *Regional Environmental Change* 19 (2019): 1549–60.

17. For some examples of an individual scientist's climate observations, see the Franklin B. Hough Papers, New York State Library, box 4. For an example of a climate change report derived from Signal Corps station data, see Adolphus Greely's *Report of Rainfall in Washington Territory, Oregon, California, Idaho, Nevada, Utah, Arizona, Colorado, Wyoming, New Mexico, Indian Territory, and Texas, for from Two to Forty Years,* H.R. 50th Cong., 1st Sess., Ex. Doc. No. 91. Washington, DC: GPO, 1889; Deborah Coen describes a similar scientific struggle with data collection in "Climate and Circulation in Imperial Austria," *Journal of Modern History* 82 (December 2010): 848–49.

18. Watson argued that clearing forests near the Champlain Valley made the region's "climate more drouthy." Winslow C. Watson, "Forests—Their Influence, Uses, and Reproduction," *Transactions of the New York State Agricultural Society for the Year 1865* (Albany, NY: Cornelius Wendell, 1866), 289–90.

19. See John Disturnell, *Influence of Climate* (New York: D. Van Nostrand, 1867), 105; Theodore C. Henry (of Abilene, KS), Addresses on "Kansas Stock Interests" and "Kansas Forestry" (Abilene, KS: Gazette Steam Printing Office, 1882), 2–3.

20. The "filaments" quotation is from "Letter from Dr. Franklin B. Hough, in Regard to the Effect of Forests in Increasing the Amount of Rainfall," H.R. 48th Cong., 2nd Sess., Ex. Doc. (Washington, DC: GPO, 1885), appendix 14, 130. For the agricultural argument, see Samuel Aughey, *The Physical Geography and Geology of Nebraska* (Omaha, NE: Daily Republican Book and Job Office, 1880), 44–45; Wilber, *Great Valleys and Prairies,* 70.

21. For an example of fire-based climate influence arguments see, for example, John Trowbridge, "Great Fires and Rain-Storms," *Popular Science Monthly,* December 1872. For ponds, see J. R. Sage, "Influence of Forests on Climate in Iowa," in the "Current Notes" section of the *American Meteorological Journal* 10, no. 14 (March 1894). For electricity and railroads, see Elliott, *Notes Taken in Sixty Years,* 309.

22. John Wesley Powell, "Our Recent Floods," *North American Review* 155, no. 429 (August 1892): 152–53. Powell's erstwhile assistant Grove Karl Gilbert held similar beliefs about uncertainty and the unknown. In a chapter titled "Water Supply," which Gilbert wrote as part of Powell's *Report on The Lands of the Arid Region of the United States* (1878; repr., Cambridge, MA: Belknap Press, 1962), Gilbert addressed issues of uncertainty and anthropogenic climate change. He argued that people should avoid forming conclusions about climate change as long as knowledge of the forces shaping climate remained in the realm of the "unknown." See Gilbert in Powell, *Report on the Lands,* 94.

23. For another example of this argument, see Samuel Temple, "Forestation and Its Discontents: The Invention of an Uncertain Landscape in Southwestern France, 1850-Present," *Environment and History* 17, no. 1 (February 2011): 15.

24. James Rodger Fleming, "Climate, Change, History," *Environment and History* 20, no. 4 (2014): 577; Mike Hulme, "Climate and Its Changes: A Cultural Appraisal," *Geo* 2, no. 1 (2015): 9.

25. Sarah Carson notes that the word "climate" "underwent considerable revision over the nineteenth and twentieth centuries." Carson, "Atmospheric Happening and Weather Reasoning: Climate History in South Asia," *History Compass* 18, no. 12 (2020).

See also James Rodger Fleming and Vladimir Jankovic, "Introduction: Revisiting *Klima*," *Osiris* 26, no. 1 (2011): 1–15.

26. Larrabee is quoted in *Popular Science Monthly* 40 (April 1892): 804.

27. Jamie L. Pietruska, "US Weather Bureau Chief Willis Moore and the Reimagination of Uncertainty in Long-Range Forecasting," *Environment and History* 17, no. 1 (2011): 79–105.

28. George E. Curtis, "The Trans-Mississippi Rainfall Problem Restated: The Rainfall in Its Relation to Kansas Farming," *American Meteorological Journal* 5, no. 2 (June 1888): 66.

29. George E. Curtis, "Analysis of the Causes of Rainfall, with Special Relation to Surface Conditions," *American Meteorological Journal* 10, no. 6 (October 1893): 274.

30. For examples of the long-running feud between Hinrichs and the meteorological establishment, see *First Biennial Report of the Central Station of the Iowa Weather Service* (Des Moines, IA: F. M. Mills, 1880) in University of Iowa—Special Collections—RG99.0039—box 2, 22–23 and Gustavus Hinrichs, "Tornadoes and Derechos," *American Meteorological Journal* 5, no. 9 (January 1889): 392–93.

31. Gustavus Hinrichs, *Second Annual Report of the Iowa State Weather Service*, printed as an appendix to the *Report of the Iowa State Agricultural Society* for the year 1877, 624.

32. For "self-registering" quotation, see Hinrichs, *First Biennial Report*, 21. For logarithm and natural laws, see Hinrichs, *Rainfall Laws Deduced from Twenty Years of Observation* (Washington, DC: Weather Bureau, 1893). "Every new law" quotation is from *Rainfall Laws*, 11.

33. The comment on the need for centuries of data to understand climate change is from Hinrichs, *First Biennial Report*, 21.

34. Gregg Crane examines James's views of certainty and uncertainty in "Playing It Safe: American Literature and the Taming of Chance," *Modern Intellectual History* 11, no. 1 (2014): 223–24. See also T. J. Jackson Lears, *Something for Nothing: Luck in America* (New York: Viking, 2003).

35. "Proper continuation" is from Hinrichs, *Second Annual Report*, 624. "Baconian" quotation is from *Rainfall Laws*, 82.

36. Lorraine Daston and Pater Galison, *Objectivity* (New York: Zone Books, 2010).

37. *Transactions of the Illinois State Horticultural Society for 1871*, proceedings of the sixteenth annual meeting held Jacksonville, December 12, 13, 14, and 15 (Chicago: Reade, Brewster, 1872). See 34 for Periam, 21 for Tice, and 193 for Baker.

38. See Maude M. Bishop, "Joseph Taplin Lovewell," *Bulletin of the Shawnee County Historical Society* 38 (December 1962).

39. For the unsettled question and "forces in nature," see J. T. Lovewell, "Kansas Meteorology," *Fourth Biennial Report of the State Board of Agriculture to the Legislature of the State of Kansas, for the Years 1883–1884* (Topeka: Kansas Publishing House: T. D. Thacher, 1885), 612–13. For continued belief in truth, see J. T. Lovewell, "Kansas Weather Service," *Transactions of the Kansas Academy of Science for 1879–1880* (Topeka: Geo. W. Martin, Kansas Publishing House, 1881), 87.

40. Lovewell, "Kansas Meteorology," 612.

41. H. K. McConnell, "Rainfalls of Kansas," *Osage County Chronicle*, March 30, 1882 (Kansas Historical Society, "Rain and Rainfall" Clippings, 551.57R).

42. "Economic Conditions Antagonistic to a Conservative Forest Policy," address delivered by B. E. Fernow (US Department of Agriculture) before the American As-

sociation for Advancement of Science, August 1892, Bernhard Fernow Papers, Cornell University, box 2, folder 23, 2.

43. See B. E. Fernow, "Introduction and Summary of Conclusions," US Department of Agriculture, Forestry Division, Bulletin 7, *Forest Influences* (Washington, DC: GPO, 1893), 9–10. Before making claims about forest influences on climate, Fernow states that "from the complication of causes which produce climatic conditions it has always been difficult to prove, when changes in a given region were observed, that they are permanent and not due merely to the general periodic variations which have been noted in all climates of the earth."

44. "Whatever the truth" is from Fernow, "The Forest as a Condition of Culture," 13. "Needful regulators" is from Fernow, "Economic Conditions," 1–2.

45. Thomas P. Roberts, "Relation of Forests to Floods," in *Proceedings of the American Forestry Congress at Its Meeting Held in Boston, September, 1885* (Washington, DC: Judd and Detweiler, 1886), 101.

46. H. R. [Hugh Rankin] Hilton, "Effects of Civilization on the Climate and Rain Supply of Kansas," a Lecture Delivered by H. R. Hilton, Esq., before the Scientific Club Of Topeka, Wednesday Evening, March 31st, 1880 (Spencer Library, University of Kansas Archives, RH C4318), 3–5.

47. *Transactions of the Illinois State Horticultural Society for 1870* (Chicago: Dunlop, Reade, and Brewster, 1871), 177 for "feeble endeavors," 197 for Tice quotation.

48. J. T. Lovewell, "Human Agency in Changing or Modifying Climate," *Quarterly Report of the Kansas State Board of Agriculture for the Quarter Ending March 31, 1892* (Topeka, KS: Hamilton Printing Company, 1892), 139–43.

49. Lovewell, "Kansas Meteorology," 140.

50. Thomas in Ferdinand V. Hayden, *The First, Second, and Third Annual Reports of the U.S. Geological Survey of the Territories for the Years 1867, 1868, and 1869 under the Department of the Interior* (Washington, DC: GPO, 1873), 236–37.

51. *Dodge City (KS) Times*, January 19, 1878.

52. Hiram Adolphus Cutting, *Forests of Vermont* (Montpelier: Vermont Watchman and State Journal Press, 1886), 3–5.

53. F. Hawn, "Source of Rains in Kansas," *Quarterly Report of the Kansas State Board of Agriculture for the Quarter Ending September 30, 1881* (Topeka: Kansas Publishing House, 1881), see 45 for "climatic relations" and 48 for "radical changes." The Sierra Nevada studies cited by Hawn were carried out by a researcher referred to as "Professor Legate." These experiments are described in Henry G. Vennor, *Vennor's Almanac and Weather Record for 1878–1879* (Montreal: Witness Printing House, 1879).

54. See Ramachandra Guha, *Environmentalism: A Global History* (New York: Longman, 2000), 30.

55. Legate's views are also summarized in a piece from the *Virginia City (NV) Enterprise* reproduced in the *Pacific Rural Press* 20, no. 19 (November 6, 1880): 291.

56. For a study of restoration theory in the nineteenth and early twentieth centuries, see Marcus Hall, *Earth Repair: A Transatlantic History of Environmental Restoration* (Charlottesville: University of Virginia Press, 2005). See also David Lowenthal, *George Perkins Marsh: Prophet of Conservation* (Seattle: University of Washington Press, 2000); Philip J. Pauly, *Fruits and Plains: The Horticultural Transformation of America* (Cambridge, MA: Harvard University Press, 2007), 80–96.

57. Cutting, *Forests of Vermont*, see 3 for a critique of profit seeking and 12 for faith in "moneymaking."

58. Marshall Berman, *All That Is Solid Melts into Air: The Experience of Modernity* (New York: Penguin Books, 1982), 95. See also Kevin Rozario, *The Culture of Calamity: Disaster and the Making of Modern America* (Chicago: University of Chicago Press, 2007). Rozario analyzes the "catastrophic logic of modernity" and the "evolving relationship" between modernity, capitalist development, and catastrophe (10–12).

59. For pro-railroad climate theory, see Elliott, *Notes*, 324. For antirailroad sentiment, see "Forest Circular No. 3," issued by Franklin B. Hough in 1877, box 40, folder 1, Franklin B. Hough Papers, New York State Library, and also J. Sterling Morton, *"Arbor Day," in Proceedings of the American Forestry Congress* (1885), 51.

60. J. L. Budd [Iowa Board of Forestry], "Possible Modification of Our Prairie Climate," in Preliminary Newspaper Report, *Sixth Annual Meeting of the American Forestry Congress*, Held in Springfield, IL, 1887 (Springfield, IL: State Register Book and Job Print, 1887), 21. For an example of climate theorists exalting the plow, see Wilber, *Great Valleys*, 70.

61. I am drawing from a rich vein of scholarship on uncertainty, especially in the history of science and science and technology studies. See Susan Leigh Star, "Scientific Work and Uncertainty," *Social Studies of Science* 15, no. 3 (August 1985): 391–427; Brian L. Campbell, "Uncertainty as Symbolic Action in Disputes among Experts," *Social Studies of Science* 15, no. 3 (August 1985): 429–53; Trevor Pinch, "The Sun-Set: The Presentation of Certainty in Scientific Life," *Social Studies of Science* 11, no. 1 (February 1981): 131–58; Michelle Murphy, *Sick Building Syndrome and the Problem of Uncertainty* (Durham, NC: Duke University Press, 2006). The field of agnotology has also yielded some important insights about uncertainty. See Robert N. Proctor, *Cancer Wars* (New York: Basic Books, 1995); Londa Schiebinger, "Agnotology and Exotic Abortifacients: The Cultural Production of Ignorance in the Eighteenth-Century World," *Proceedings of the American Philosophical Society* 149, no. 3 (September 2005): 316–43.

62. For a discussion of "reducible" and "irreducible uncertainty," see Phaedra Daipha, "Weathering Risk: Uncertainty, Weather Forecasting, and Expertise," *Sociology Compass* 6, no. 1 (2012): 16. Whereas reducible uncertainty can be managed and sometimes calculated and quantified, irreducible uncertainty presents a more intractable problem because of its "aleatory and stochastic" nature. See also Daipha, *Masters of Uncertainty: Weather Forecasters and the Quest for Ground Truth* (Chicago: University of Chicago Press, 2015).

63. Climate change debates focused especially on the West, Midwest, and Northeast, but climate perceptions in the US South certainly influenced the American cultural and scientific imagination throughout the long nineteenth century. See Jason Hauser, "'Scarce Fit for Anything but Slaves and Brutes': Climate in the Old Southwest, 1798–1855," *Alabama Review* 70, no. 1 (April 2017).

64. For climatology as a "professional/institutional project," see Zeke Baker, "Agricultural Capitalism, Climatology and the 'Stabilization' of Climate in the United States, 1850–1920," *British Journal of Sociology* 72 (June 2020): 1–18. See also Jeremy Vetter, "Knowing the Great Plains Weather: Field Life and Lay Participation on the American Frontier during the Railroad Era," *East Asian Science, Technology, and Society: An International Journal* 13, no. 2 (2019): 195–213.

65. The *American Meteorological Journal* was the most prestigious Gilded Age publication dealing with climatological and meteorological questions. It published innumerable articles on anthropogenic climate change, beginning with S. R. Thompson's 1884 piece "The Rainfall of Nebraska." *American Meteorological Journal* 2, no. 1 (June 1884).

For examples of nuanced data discussions in popular publications, see G. E. Tewksbury, *The Kansas Picture Book* (Topeka, KS: A. S. Johnson, 1883), 34, and Henry Inman's article from the March 12, 1879, *Chronoscope* (Larned, KS). Inman, for instance, takes issue with methods used for measuring "aqueous precipitation of the Central Great Plains."

66. The quotation about "corporate interests" was written by the editors of the journal *Science*, who also complained that the debate "has not been of a purely scientific character." See "The Influence of Forests on the Quantity and Frequency of Rainfall," *Science* 12, no. 303 (November 23, 1888): 241.

67. Sources published about 1890, for example, indicate that climate theories remained as eclectic as in earlier years. J. R. Sage, the director of the Iowa Weather and Crop Service, sought to refute claims that draining bogs and wetlands could cause climatic desiccation. He claimed that "the records show no evidence whatever of a change of climate in this western region" and denounced all "gloomy prognostication." Just a few months earlier, however, L. F. Andrews, assistant secretary of the Iowa Board of Health, wrote a paper in the same publication claiming that Iowa's climate could "most assuredly" be "modified" for the worse and for the better by human influences. Andrews lamented that the "greed of pelf and the aggrandizement of the aristocracy have wrought desolation as cheerless, and complete as that of the moon." See J. R. Sage, "The Practical Value of Reliable Crop and Weather Reports," *Monthly Review of the Iowa Weather and Crop Service, Co-operating with the United States Signal Service*, November 1890, 3; L. F. Andrews, "The State of Iowa: Its Topography, Drainage, Fertility, Climate, and Healthfulness," *Monthly Review of the Iowa Weather and Crop Service, Co-operating with the United States Signal Service*, April and May 1890, 6–7.

68. Paul Edwards, *A Vast Machine: Computer Models, Climate Data, and the Politics of Global Warming* (Cambridge, MA: MIT Press, 2010).

69. See, for example, Joshua Howe, *Behind the Curve: Science and the Politics of Global Warming* (Seattle: University of Washington Press, 2014).

70. For some examples of more recent research on the influence of forests on climate, see J. G. Canadell and M. R. Raupach, "Managing Forests for Climate Change Mitigation," *Science* (AAAS) 320 (June 13, 2008): 1456–57; C. Streck and S. M. Scholz, "The Role of Forests in Global Climate Change: Whence We Come and Where We Go," *International Affairs* 82 (October 4, 2006): 861–79; David K. Adams et al., "The Amazon Dense GNSS Meteorological Network: A New Approach for Examining Water Vapor and Deep Convection in the Tropics," *Bulletin of the American Meteorological Society* 96 (December 2015): 2151–65; Mingfang Zhang and Xiaohua Wei, "Deforestation, Forestation, and Water Supply," *Science* 371 (March 5, 2021): 990–91. See also Sid Perkins, "Crop Irrigation Could Be Cooling Midwest," *Science News*, January 22, 2010.

71. Naomi Oreskes and Erik Conway, *Merchants of Doubt: How a Handful of Scientists Obscured the Truth on Issues from Tobacco Smoke to Global Warming* (New York: Bloomsbury Press, 2010).

72. For "strategic positivism" see Bruno Latour, "Telling Friends from Foes in the Time of the Anthropocene," in *The Anthropocene and the Global Environmental Crisis: Rethinking Modernity in a New Epoch*, ed. Clive Hamilton, Christophe Bonneuil, and François Gemenne (London: Routledge, 2015). Despite his recent suggestion to occasionally employ "strategic positivism," Latour has also called for us to "suspend our certainties." Latour, *The Politics of Nature: How to Bring the Sciences into Democracy*, trans. Catherine Porter (Cambridge, MA: Harvard University Press, 2004), 21 and 25. For an overview of "strategic positivism," a concept inspired by Gayatri Chakravorty Spivak's

"strategic essentialism," see Elvin Wyly, "Strategic Positivism," *Professional Geographer* 61, no. 3 (2009): 310–22.

Chapter 1

1. *Topeka (KS) Weekly Tribune*, March 27, 1879, Kansas State Historical Society, "Rain and Rainfall" clippings, 551.57R. For more on Joseph Henry, see James Rodger Fleming, *Meteorology in America, 1800–1870* (Baltimore: Johns Hopkins University Press, 1990), 127–28. It is unclear which map Henry is describing in this passage, but it may have resembled C. S. Sargent's forestry map or Charles Schott's rainfall charts (these maps will be discussed in chapter 4).

2. For a discussion of the intertwined nature of climatic and cultural debates, see Lawrence Culver, "Seeing Climate through Culture," *Environmental History* 19, no. 2 (April 2014): 311–18.

3. Frederic Hawn, "Climatic Changes," *Western Homestead* (Leavenworth, KS) 3, no. 3 (August 1880): 40. For a biographical sketch of Frederic Hawn, see "Panorama of the Many Industries of Leavenworth," *Leavenworth Times*, September 23, 1887. Born in Danube, New York, in 1810, Hawn traveled throughout the West from the 1850s until the 1870s, working as a civil engineer, land surveyor, and coal mine operator. He endured several periods of "disheartening struggles" over the course of his travels. Perhaps these difficulties and failures made him sensitive to the plight of struggling farmers in the Great Plains and Intermountain West.

4. *Report of Rainfall in Washington Territory, Oregon, California, Idaho, Nevada, Utah, Arizona, Colorado, Wyoming, New Mexico, Indian Territory, and Texas, for from Two to Forty Years*, H.R. 50th Cong., 1st Sess., Ex. Doc. No. 91 (Washington, DC: GPO, 1889), 9.

5. *Sixth Annual Meeting of the American Forestry Congress*, held in Springfield, Illinois, 1887 (Springfield: State Register Book and Job Print, 1887), 23 and 26, from Forest History Society Library, George H. Wirt Collection. Miller was not present at the conference, but other members of the Congress read his paper. Several other sources echo Miller's view of Native Americans as careful users of fire. J. K. Macomber, for example, took issue with the notion that Native Americans carelessly set fires. In an 1880 paper discussing the cause of the Plains' lack of trees, Macomber argued that Iowa's relative lack of timber could not be attributed to fires ignited by Native Americans. See J. K. Macomber, "Adaptability of Prairie Soils for Timber Growth," *Transactions of the Iowa State Horticultural Society for 1879* (Des Moines, IA: F. M. Mills, 1880), 293.

6. See, for example, Steven Conn, *History's Shadow: Native Americans and Historical Consciousness in the Nineteenth Century* (Chicago: University of Chicago Press, 2004), 76–77.

7. Edwin A. Curley, *Nebraska 1875: Its Advantages, Resources, and Drawbacks* (1875; repr., Lincoln: University of Nebraska Press, 2007), xiv, xx, "Preface," 107.

8. Elliott West, *The Contested Plains: Indians, Goldseekers and the Rush to Colorado* (Lawrence: University of Kansas Press, 1998), especially 190–91.

9. For a discussion of nineteenth-century writings depicting Native Americans as poor stewards of the land, see Shepard Krech, *The Ecological Indian: Myth and History* (New York: W. W. Norton, 1999), 101–5. Krech focuses especially on the notion that Native Americans set fires indiscriminately. Diana K. Davis's *The Arid Lands: History, Power, Knowledge* (Cambridge, MA: MIT Press, 2016) details how colonial settlement

projects in arid regions throughout the globe have used degradation and desiccation narratives as weapons against "indigenous, local populations" (82).

10. Transcript of President Andrew Jackson's message to Congress on "Indian Removal" (1830), Ourdocuments.gov, https://www.ourdocuments.gov/doc.php?flash=false&doc=25&page=transcript (accessed January 5, 2021).

11. Nathan Meeker, in *Report of the Public Lands Commission Created by the Act of March 3, 1879, Relating to the Public Lands in the Western Portion of the United States and to the Operation of Existing Land Laws* (Washington, DC: GPO, 1880), 294. Meeker was only a lukewarm supporter of climate change theories.

12. Charles Dana Wilber, *The Great Valleys and Prairies of Nebraska and the North-west* (Omaha, NE: Daily Republican Print, 1881), 70.

13. *Topeka (KS) Commonwealth*, pamphlet, "One Hundredth Meridian" (1879), Kansas Historical Society.

14. Conn, *History's Shadow*, 70 for "part of nature," 215 for narratives, 199 for "ahistorical."

15. Ferdinand V. Hayden, *The Great West* (Bloomington, IL: Charles R. Brodix, 1880), 5 for "Wild Indians" (this section was probably written by the publishers) and 67–69 for the Colorado Plateau. Hayden collaborated closely with economic interests, leading many to label him a booster-surveyor. He was far more inclined to support unmitigated agricultural and railroad-driven expansion than his colleague and competitor John Wesley Powell. See Mike Foster, *Strange Genius: The Life of Ferdinand Vandeveer Hayden* (Niwot, CO: Roberts Rinehard, 1994).

16. J. W. Powell articulated the "whence came" question. See Conn, *History's Shadow*, especially 106, 122, 133.

17. Mark Spence discusses George Bird Grinnell and other figures who simultaneously "bemoaned the destruction of native societies" and sought to "civilize" and marginalize them. See Mark Spence, *Dispossessing the Wilderness: Indian Removal and the Making of the National Parks* (New York: Oxford University Press, 1999), 78. See also Miles Powell, *Vanishing America: Species Extinction, Racial Peril, and the Origins of Conservation* (Cambridge, MA: Harvard University Press, 2016).

18. C. B. Boynton and T. B. Mason, *A Journey through Kansas; with Sketches of Nebraska: Describing the Country, Climate, Soil, Mineral, Manufacturing, and Other Resources* (Cincinnati, OH: Moore, Wilstach, Keys, 1855), 89–91.

19. Jason Colavito, *The Mound Builder Myth: Fake History and the Hunt for a "Lost White Race"* (Norman: University of Oklahoma Press, 2020).

20. J. H. Tice, "Meteorological Effects of Forests," *Transactions of the Illinois State Horticultural Society for 1870*, Held at Galesburg, December 13, 14, 15, and 16 (Chicago: Dunlop, Reade, and Brewster, 1871), 166–75.

21. Mark W. Harrington, "Climate of Santa Fe," *American Meteorological Journal* 2, no. 3 (July 1885): 120; "Is Our Average Rainfall Diminishing?" *Scientific American* 54, no. 7 (February 13, 1886): 96.

22. Shephard Krech has discussed the long-running debate about the meaning of the Hohokam. See Krech, *Ecological Indian*, 45–68.

23. Franklin B. Hough, *The Elements of Forestry* (Cincinnati, OH: Robert Clarke, 1882), 27.

24. Patrick Hamilton, *The Resources of Arizona* (San Francisco, CA: Bancroft, 1883), 7 for images of a "Golden Age," 144 for restoration of former splendor.

25. W. A. Glassford, "Charts and Tables of Rain-Fall on the Pacific Slope," in *Report*

of Rainfall in Washington Territory, Oregon, California, Idaho, Nevada, Utah, Arizona, Colorado, Wyoming, New Mexico, Indian Territory, and Texas, for from Two to Forty Years, US Signal Corps. H.R. 50th Cong., 1st Sess., Ex. Doc. No. 91 (Washington, DC: GPO, 1889), 28.

26. W. A. Glassford, "Climate of New Mexico," in "Report on the Climatology of the Arid Region of the United States, with Reference to Irrigation," by Adolphus Greely, in *Report on the Climate of New Mexico,* by the Chief Signal Officer of the Signal Corps (Washington, DC: GPO, 1891), 23–24. Like many of his contemporaries, Glassford admired Native American societies from the past. In an 1890 report on the climate of Arizona, he wrote that "in the Salt River Valley, the Mormons . . . owe their prosperity to the aqueduct system and this in turn they owe in large part to the labors of a former race of whom all knowledge has vanished." See W. A. Glassford. "Climate of Arizona, with Particular Reference to the Rainfall and Temperature, and Their Influence upon the Irrigation Problems of the Territory," in US Signal Office, *Irrigation and Water Storage in the Arid Regions* (Washington, DC: GPO, 1891), 309.

27. Eugene W. Hilgard discusses the Powell Survey's map in "Climates of the Pacific Slope," from the *Report Made under the Direction the Commissioner of Agriculture,* by E. W. Hilgard, T. C. Jones, and R. W. Furnas (Washington, DC: GPO, 1882), 16–18, Nebraska State Historical Society, J. Sterling Morton Pamphlet Collection, vol. 1.

28. G. K. Gilbert, in John Wesley Powell, *Report on The Lands of the Arid Region of the United States* (1878; repr., Cambridge, MA: Belknap Press, 1962), 84–85.

29. Richard Francaviglia, *The Mapmakers of New Zion: A Cartographic History of Mormonism* (Salt Lake City: University of Utah Press, 2015), 120. Historians including Donald Worster, Jared Farmer, and Francaviglia have challenged the idyllic narrative of the Mormon transformation of Utah from desert into garden. See Richard Francaviglia, *The Mormon Landscape: Existence, Creation, and Perception of a Unique Image of the American West* (New York: AMS Press, 1978), 86; Jared Farmer, *On Zion's Mount: Mormons, Indians, and the American Landscape* (Cambridge, MA: Harvard University Press, 2008), 105; Donald Worster, *Rivers of Empire: Water, Aridity, and the Growth of the American West* (New York: Pantheon Books, 1985), 74–80.

30. Jules Remy and Julius Brenchley, *A Journey to Great-Salt-Lake-City* (London: W. Jeffs, 1861), iii for "inaccuracies," 268–69 for "fraud" and "mysticism," and 152 for a discussion of climate and lake level.

31. Remy and Brenchley, *Journey to Great-Salt-Lake-City,* 152. Richard F. Burton, another traveler to Utah, asserted that increased rainfall levels in Utah "may be attributed to cultivation and plantation; thus also may be explained the N.A. Indian's saying that the pale-face brings with him the rain." Richard F. Burton, *The City of the Saints: Across the Rocky Mountains to California* (New York: Knopf, 1963), 305–6. Burton traveled to Utah in 1860; his narrative was originally published in 1861 and reprinted in 1881.

32. Samuel Bowles, *Across the Continent: A Summer's Journey to the Rocky Mountains, the Mormons, and the Pacific States* (Springfield, MA: Samuel Bowles, 1865), vi, 19, 89, 132. For more on Bowles's narrative, see Katrine Quinn, "'Across the Continent . . . and Still the Republic!' Inscribing Nationhood in Samuel Bowles's Newspaper Letters of 1865," *American Journalism* 31, no. 4 (2014).

33. Farmer, *On Zion's Mount,* 105, 126–28.

34. F[loyd] P[erry] Baker, *Preliminary Report on the Forestry of the Mississippi Valley and Tree Planting on the Plains* (Washington, DC: GPO, 1883), 25.

35. W[illiam] B[abcock] Hazen, "The Great Middle Region of the United States,

and Its Limited Space of Arable Land," *North American Review* 120, no. 246 (January 1875): 18, 23.

36. *Report of the Public Lands Commission Created by the Act of March 3, 1879*, 493.

37. F. V. Hayden, *Preliminary Report of the United States Geological Survey of Wyoming and Portions of Contiguous Territories* (Washington, DC: GPO, 1871), 456.

38. Hayden, *Great West*, 1880, 324. Hayden expressed his relief that Mormon excesses had been brought under control by the influence of non-Mormon "gentiles" in Utah: "Mormonism to-day is a different thing from that of Brigham Young's time. . . . Gradually that wildness of fanaticism died out through contact with Gentiles and other causes" (326–27).

39. Greely, *Report of Rainfall*, 11.

40. Gilbert, in Powell, *Report on the Lands of the Arid Region*, 84.

41. Utah Board of Trade, *Resources and Attractions of the Territory of Utah* (Omaha, NE: Omaha Republican Publishing House, 1879), 8 and 12. Whereas Great Plains boosters pointed to Utah settlers as climate improvement pioneers, some sources invoked Kansas and Nebraska as leading examples of climatic amelioration. The Utah Board of Trade wrote that the "increased humidity has followed the settlement and cultivation of the Mississippi Valley Prairies, and it is not unlikely that it is doing so in Utah, although there is not sufficient data as yet upon which to assert it."

42. Gregg Mitman, "Geographies of Hope: Mining the Frontiers of Health in Denver and Beyond," *Osiris* 19 (2004): 93–111.

43. Utah Board of Trade, *Resources and Attractions*, 14–15. For a discussion of Utah's tourism industry and notions of "geographical medicine," see Farmer, *On Zion's Mount*, 107–14. Health tourism–based opposition to the implementation of climate improvement schemes highlights divisions among those who believed in anthropogenic climate change. Edward Powers, who endorsed the notion that "rain can be produced by human agency," speculated that California and Colorado "might prefer to retain their character as great sanatariums for the cure of pulmonary complaints, rather than gain the advantage to agriculture which frequent summer rains would give." Powers gained widespread renown for his arguments about war and cannonade-induced rainfall. See Edward Powers, *War and the Weather* (1871; repr., Delevan, WI: Edward Powers, 1890), 5, 116–17.

44. *Nebraska Farmer* 1, no. 9 (September 1877): 4.

45. "Professor Aughey Lived a Useful Life: Notes of the Address Delivered by Rev. W. G. M. Hayes at the Obsequies of the Late Professor Samuel Aughey" (1912)—University of Nebraska Special Collections RG52/01, box no. 13. Hayes describes Aughey as a "man of recognized standing as a scientist who was also a theologian." He quotes Aughey as saying there should not be "any contradiction" between science and religion. Gustavus Hinrichs also wrote about the importance of reconciling science and faith. See "Faith and Science," a "Lecture Delivered by Prof. Gustavus Hinrichs before the students of the [Iowa] State University." Although undated, the lecture is in a folder with papers from 1867, so it can be presumed to be from that year. University of Iowa Library, University Archives, RG 99.0039. G. Hinrichs Papers, box 1.

46. Over the course of the nineteenth century, thinkers ranging from Thomas Cole to John Muir used notions of sin and atonement in crafting narratives about environmental degradation and recovery. For a discussion of the role of religion in shaping American ideas about landscape and wilderness, see Mark Stoll, "Religion 'Irradiates' the Wilderness," in *American Wilderness: A New History*, ed. Michael Lewis (Oxford: Oxford University Press, 2007), 44–45; Stoll, *Inherit the Holy Mountain: Religion and the Rise of American Environmentalism* (Oxford: Oxford University Press, 2015).

Chapter 2

1. For histories of Gilded Age capitalism and political economy, see Alan Trachtenberg, *The Incorporation of America: Culture and Society in the Gilded Age* (New York: Hill and Wang, 1982); Rosanne Currarino, *The Labor Question in America: Economic Democracy in the Gilded Age* (Urbana: University of Illinois Press, 2011); Richard White, *The Republic for Which It Stands: The United States during Reconstruction and the Gilded Age* (New York: Oxford University Press, 2017), especially 213–52; Noam Maggor, *Brahmin Capitalism: Frontiers of Wealth and Populism in America's First Gilded Age* (Cambridge, MA: Harvard University Press, 2017).

2. For "moral complexity," see Patricia Nelson Limerick, *The Legacy of Conquest: The Unbroken Past of the American West* (New York: Norton, 1987), 54. See also Karl Jacoby, *Shadows at Dawn: An Apache Massacre and the Violence of History* (New York: Penguin, 2009), 6; Douglas Sackman, *Wild Men: Ishi and Kroeber in the Wilderness of Modern America* (Oxford: Oxford University Press, 2010), especially 10, 115, 314.

3. Richard Grove described the appeal of environmental redemption in his study of colonial conservation. According to Grove, conservation in tropical Edens "offered the possibility of redemption" during a time of "great uncertainty" about "the long-term security of colonial rule." See Grove, *Green Imperialism: Colonial Expansion, Tropical Island Edens and the Origins of Environmentalism, 1600–1860* (Cambridge: Cambridge University Press, 1995), 6 for "uncertainty," 13 for "redemption."

4. Elbridge Gale, "Forest Tree Culture," paper read before the Kansas State Horticultural Society, Manhattan, December (?) 7, 1878, 14–18. For more on Gale and Kansas forestry in general, see Brian Allen Drake, "Waving 'a Bough of Challenge': Forestry on the Kansas Grasslands," *Great Plains Quarterly* 23, no. 1 (Winter 2003): 19–34. Drake characterizes Gale's tone as "evangelical" (23).

5. Craig Miner, *West of Wichita: Settling the High Plains of Kansas, 1865–1890* (Lawrence: University of Kansas Press, 1986), 3.

6. For a discussion of failure and failure-related anxieties in American culture, see Scott Sandage, *Born Losers: A History of Failure in America* (Cambridge, MA: Harvard University Press, 2005).

7. Ellwood Cooper, *Forest Culture and Eucalyptus Trees* (San Francisco, CA: Cubery, 1876), 10, 13–14, 19.

8. M. C. Read, "The Preservation of Forests on the Headwaters of Streams" and "The Proper Value and Management of Government Timber Lands and the Distribution of North American Forest Trees," papers read at the United States Department of Agriculture, May 7–8, 1884, Department of Agriculture, Miscellaneous, Special Report no. 5 (Washington, DC: GPO, 1884); see 28 for "personal greed," 37 for "arid lands." In denouncing personal greed, Read aligned himself against the powerful narrative about individualism and its role in the conquest of the West. Alan Trachtenberg discusses how the cultural notion of the individualist settler helped justify the industrialization of the West by corporate interests. See Alan Trachtenberg, *The Incorporation of America: Culture and Society in the Gilded Age* (New York: Hill and Wang, 1982), 24. Similarly, Richard White, Donald Worster, and many others have pointed out that the concept of the West as a "bastion of individualism," though deeply entrenched, overlooks the role that the federal government and large corporations played in the development of the West. See Richard White, *Railroaded: The Transcontinentals and the Making of Modern America* (New York: Norton, 2011), xxii; Donald Worster, *Rivers of Empire: Water, Aridity, and the Growth of the American West* (New York: Pantheon Books, 1985).

9. In *Rivers of Empire*, Worster demonstrates how large-scale irrigation and reclamation in the West enriched elites at the expense of ordinary people. Similarly, many forest- planting initiatives such as those supported by the Timber Culture Act devolved into land speculation despite their lofty egalitarian intentions. See C. Barron McIntosh, "Use and Abuse of the Timber Culture Act," *Annals of the Association of American Geographers* 65, no. 3 (September 1975).

10. F. W. Hart, "Report on Forestry—Needs of Our Prairie State," *Transactions of the Iowa State Horticultural Society for 1879* (Des Moines, IA: F. M. Mills, 1880), 274–76. As proof "that forests induce additional rainfall," Hart cites Gustavus Hinrichs's maps showing "that the lines of equal amount of rainfall correspond very nearly with those of equal quantity of woodlands."

11. Martin Mohler, "Kansas Agriculture, Prospectively Considered," in *Report of the Kansas State Board of Agriculture for the Quarter Ending March 31, 1888* (Topeka: Kansas Publishing House, 1888), 10.

12. Rodney Welch, "How the West Has Moved On," address delivered at Lincoln, September 27, 1877, during the Nebraska State Fair, Nebraska State Board of Agriculture, 1877, Nebraska State Historical Society, J. Sterling Morton Pamphlet Collection, vol. 4, no. 9.

13. Aughey is quoted in the *Nebraska Farmer* 1, no. 10, October 1887. The phrase "vandal hand of man" seems to have been in wide circulation among climate theorists. Cutting describes the "vandal spirit in man" in his *Lectures on Plants, Fertilization, Insects, Forestry, Farm Homes, Etc.* (Montpelier, VT: Foreman, 1882), 79–80.

14. For "speculators" and "shiftless people," see W. A. Yingling, *Westward, or Central-Western Kansas* (Ness City, KS: Star Printing, 1890); Kansas State Historical Society, 917.81, Pam vol. 1, nos. 16, 13 and 19. For "nomads," see "Greenwood County, Kansas," an undated pamphlet meant for immigrants, University of Kansas Special Collections, Spencer library—the latest date mentioned in the pamphlet is 1884, so presumably the pamphlet dates from soon thereafter.

15. In 1882, for example, F. P. Baker argued that the "original forest lands of Colorado are being converted into deserts." See Baker, *Preliminary Report*, 7.

16. In 1879, Henry Inman envisioned a future in which the East had been "rendered almost uninhabitable through the wantonness of man." See the *Larned (KS) Chronoscope*, March 12, 1879, Kansas State Historical Society, "Rain and Rainfall" Clippings 551.57R.

17. Charles Dana Wilber, *The Great Valleys and Prairies of Nebraska and the Northwest* (Omaha, NE: Daily Republican Print, 1881), 70.

18. Jonathan Periam, "Forest Tree Planting, as a Means of Wealth," *Transactions of the Illinois State Horticultural Society for 1871* (Chicago: Reade, Brewster, 1872), 37. Iowa horticulturists voiced similar concerns about the negative influence of agriculture and settlement on grasses and climate. See, for example, the testimony of "Mr. Dixon" in *Transactions of the Iowa State Horticultural Society for 1879* (Des Moines, IA: F. M. Mills, 1880), 280.

19. Daniel Burge has shown that opposition to Manifest Destiny was far more widespread than most historians have acknowledged. Although his article focuses on the antebellum period, Burge details the often humorous strategies employed by critics who challenged the core of expansionist ideology. See Daniel Burge, "Manifest Mirth: The Humorous Critique of Manifest Destiny, 1846–1858," *Western Historical Quarterly* 47 (Autumn 2016): 283–302, especially 284–86.

20. Marvin E. Kroeker, *Great Plains Command: William B. Hazen in the Frontier West* (Norman: University of Oklahoma Press, 1976).

21. W. B. Hazen, *Our Barren Lands: The Interior of the United States West of the One-Hundredth Meridian and East of the Sierra Nevada* (Cincinnati, OH: R. Clarke, 1875), 8–10.

22. Hazen and Custer had a long and contentious relationship—Hazen criticized Custer's book *My Life on the Plains* and his conduct over the course of his Plains campaigns against Native Americans. See Marvin E. Kroeker, "Deceit about the Garden: Hazen, Custer, and the Arid Lands Controversy," *North Dakota Quarterly* 38 (Summer 1970): 5–21. The controversy between Hazen and Custer attracted significant press attention. Several newspapers, including the *Minneapolis Tribune*, sided with Custer and the railroads. See Hazen, *Our Barren Lands*, 13.

23. Hazen, *Our Barren Lands*, 10–11 for "deceived," 35 for Pacific-based climate patterns, 48 for afforestation.

24. Hazen, *Our Barren Lands*, 51. White describes the "social failure" that accompanied the building of the transcontinental lines and argues that "railroads and the modern state were co-productive." See White, *Railroaded*, xxvi and 511.

25. In addition to Elliott's famous Kansas Pacific Railroad–sponsored plantations in western Kansas, railroads supported 1870s Kansas experimental plantations by Louis Watson and John Bonnell. See Kansas State Historical Society, P. J. Jennings Papers, MS Coll. 404, box 1, folder titled "Watson, Dr. Louis, correspondence, 1865–875."

26. Sidney Dillon [president of the Union Pacific Railway Company], "The West and the Railroads," *North American Review* 152, no. 413 (April 1891): 445, 448. Charles Francis Adams, another railroad magnate, advanced a similar argument in "The Rainfall on the Plains," *Nation*, November 14, 1887, 417. Some climate theorists rushed to the defense of railroads and controversial railroad land grants. In 1871 Clinton C. Hutchinson, a Kansan who supported agriculturally induced climate change, accused railroad opponents of being demagogues and also criticized eastern "associations representing the laboring classes" for attacking railroad land grants as antidemocratic and monopolist. Hutchinson wrote, "It is nonsense to say the necessary roads will be built without land grants." See C. C. Hutchinson, *Resources of Kansas: Fifteen Years Experience* (Topeka, KS: Author, 1871), 212–13.

27. Hamilton, *Resources of Arizona*, 15–16.

28. Richard Smith Elliott, *Notes Taken in Sixty Years* (St. Louis, MO: R. P. Studley, 1883), 324.

29. Henry Allen Hazen, "Variation of Rainfall West of the Mississippi River," *Signal Service Notes 7*, US War Department (Washington, DC: Office of the Chief Signal Officer, 1883), 3.

30. "The Influence of Forests on the Quantity and Frequency of Rainfall," *Science* 12, no. 303 (November 23, 1888): 241.

31. Donald Worster's and Richard White's works on western and Gilded Age history tend to either dismiss climate theory or cite it as evidence of Gilded Age hubris and railroad rhetoric. See, for example, Donald Worster, *Dust Bowl: The Southern Plains in the 1930s* (Oxford: Oxford University Press, 1979), 81–82; White, *Railroaded*, 487–88. White's statement that climate change theories "were not science" assumes a narrow and teleological definition of "science" while eliding the climate writings of authors who were not part of the booster movement.

32. Julius Sterling Morton, "Arbor Day," in *Proceedings of the American Forestry Con-*

gress at Its Meeting Held in Boston, September, 1885 (Washington, DC: Judd and Detweiler, 1886), 51; Maximilian G. Kern, "The Relation of Railroads to Forest Supplies and Forestry, "Department of Agriculture, Forestry Division, Bulletin 1 (Washington, DC: GPO, 1887), 7.

33. Douglas Sackman describes figures who could not decide whether the railroad was an "angel or a demon." See Sackman, *Wild Men,* 131. See also Wolfgang Schivelbusch, *The Railway Journey: The Industrialization of Space and Time in the 19th Century* (Berkeley: University of California Press, 1986).

34. For more on the ranching and beef industry in the West, see William Cronon, *Nature's Metropolis: Chicago and the Great West* (New York: W. W. Norton, 1991), 207–59; Joshua Specht, "The Rise, Fall, and Rebirth of the Texas Longhorn: An Evolutionary History," *Environmental History* 21, no. 2 (April 2016): 343–63; Specht, *Red Meat Republic* (Princeton, NJ: Princeton University Press, 2019).

35. *Report of the Public Lands Commission,* 387.

36. McIntosh, "Use and Abuse of the Timber Culture Act."

37. Walter Kollmorgen, "The Woodsman's Assaults on the Domain of the Cattleman," *Annals of the Association of American Geographers* 59, no. 2 (June 1969): 215–39.

38. Wilber, *Great Valleys and Prairies,* 144.

39. G. E. Tewksbury, *The Kansas Picture Book* (Topeka, KS: A. S. Johnson, 1883), 19.

40. Edgar Guild, "Western Kansas: Its Geology, Climate, Natural History, Etc.," *Kansas City Review of Science and Industry,* vol. 3 (Kansas City, MO: Ramsey, Millet, and Hudson, 1885), 463, 467.

41. T. C. Henry in the *Topeka (KS) Commonwealth,* February 7, 1882.

42. Silas Bent, "Meteorology of the Mountains and Plains of North America, as Affecting the Cattle-Growing Industries of the United States," *American Meteorological Journal* 1, no. 11 (March 1885): 485.

43. Bent, "Meteorology," 481.

44. "Nearly impervious to water" is from F. Hawn, "Influence of Forests on Climate: Can the Plains Be Reclaimed by Tree-Planting?" *Kansas Magazine* 3, no. 6 (June 1873): 488.

45. See H. R. Hilton, "Moisture Economy in Kansas—Current Notes," *American Meteorological Journal* 6, no. 5 (September 1889–October 1889): 288–89.

46. Specht argues that even though the Texas longhorn was memorialized as a relic of a natural, precapitalist past, it was a technological creation and a "critical piece of a highly developed and capitalized ranching system." See Specht, "Rise, Fall, and Rebirth of the Texas Longhorn," 346.

47. For "first nature" and "second nature" in Chicago and the West, see Cronon, *Nature's Metropolis,* 56. See also Neil Smith, *Uneven Development: Nature, Capital, and the Production of Space* (Oxford: Blackwell, 1984).

48. J. H. Beadle. *The Undeveloped West, or Five Years in the Territories* (Philadelphia: National Publishing Company, 1873), 55.

49. H. R. Hilton, "Influence of Climate and Climatic Changes upon the Cattle Industry of the Plains," *Report for the Kansas State Board of Agriculture* (1888), 142.

50. Hilton, "Influence of Climate," 144–45.

51. Hilton, "Influence of Climate," 143.

52. *Letter from George B. Loving, Esq., of Fort Worth, Tex., in Regard to the Losses of Cattle during the Winter of 1884-'85, the Decline in the Value of Stock, and the Future of the Stock-Growing Interests of Texas,* H.R. Doc., 48th Cong., 2nd Sess., appendix 15 (Washington, DC: GPO, 1885), 133.

53. H. W. S. Cleveland, *Landscape Architecture as Applied to the Wants of the West* (1873; repr., Amherst: University of Massachusetts Press, 2002), 16–17, 29.

54. Cleveland, *Landscape Architecture*, 113.

55. Aaron Sachs, *Arcadian America: The Death and Life of an Environmental Tradition* (New Haven, CT: Yale University Press, 2013), 242, 246.

56. *Kansas City Journal*, December 14, 1887.

57. F. M. Clarke, "Shall We Build Reservoirs?" *Annual Report of the Colorado State Board of Horticulture and State Agricultural and Forestry Association* (1887–88), 341, 343. For other discussions of irrigation reservoir–induced climate change, see Joseph Nimmo, "Report on the Internal Commerce of the United States," Department of the Treasury, Bureau of Statistics (Washington, DC: GPO, 1885), 105; John Hay, "Atmospheric Absorption and Its Effect upon Agriculture," *Proceedings of the Eighteenth Annual Meeting of the Kansas State Board of Agriculture* (Topeka, 1890), 127–29.

58. For more on the shelterbelt program, see *Possibilities of Shelterbelt Planting in the Plains Region*, prepared under the direction of The Lake States Forest Experiment Station, USFS (Washington, DC: GPO, 1935), and Joel J. Orth, "The Conservation Landscape: Trees and Nature on the Great Plains" (PhD diss., Iowa State University, 2004).

59. See William Meyer, *Americans and Their Weather* (Oxford: Oxford University Press, 2000), and Richard E. Lingenfelter, *Death Valley and the Amargosa: A Land of Illusion* (Berkeley: University of California Press, 1986), 97. For a transnational study of similar schemes in other colonial contexts, see Meredith McKittrick, "Theories of 'Reprecipitation' and Climate Change in the Settler Colonial World," *History of Meteorology* 8 (2017): 74–94.

60. "Reports of J. E. James and Richard H. Stretch, Civil Engineers, &c., on the Practicability of Turning the Waters of the Gulf of California into the Colorado Deserts and the Death Valley," S. Doc., 43rd Cong., 1st Sess., Misc. Doc. no. 84 (March 19, 1874).

61. L. P. Brockett, *Our Western Empire, or The New West Beyond the Mississippi* (Columbus, OH: William Garretson, 1881), 83 for "diminished the rainfall," 208 for restoration, 497 for "evaporation."

62. "Reports of J. E. James and Richard H. Stretch," 4.

63. Meyer describes California-based opposition to the climate engineering proposals in *Americans and Their Weather*, 100. Meyer also cites John Wesley Powell as an opponent of the plans advanced by Brockett, Frémont, and others.

64. "Reports of J. E. James and Richard H. Stretch," 4.

65. William S. Kiser, *Coast-to-Coast Empire: Manifest Destiny and the New Mexico Borderlands* (Norman: University of Oklahoma Press, 2018), 9.

66. Samuel Truett, *Fugitive Landscapes: The Forgotten History of the U.S.–Mexico Borderlands* (New Haven, CT: Yale University Press, 2006), 9, 67, 134–48.

67. In her study on property rights and landownership in nineteenth-century New Mexico, Maria Montoya shows how Anglo-Americans "argued that both Indians and Mexicans 'wasted' land, misusing its resources. They therefore felt justified in advocating dispossession and improvement of both the land and its inhabitants." Montoya also demonstrates that "there were economic coalitions . . . that cut across racial and ethnic boundaries." See Maria Montoya, *Translating Property: The Maxwell Land Grant and the Conflict over Land in the American West* (Berkeley: University of California Press, 2002), 13–14.

68. Elias Brevoort, *New Mexico: Her Natural Resources and Attractions* (Santa Fe, NM: Elias Brevoort, 1874), ix for "the population," 15 for climate discussion, 52 for railroad construction.

69. Stephen Dorsey, "Land Stealing in Mexico, a Rejoinder," *North American Review* 145 (October 1887), 397 for land grants, 407 for the climate question.

70. Dorsey, "Land Stealing in Mexico," 406 for "ordinary farming."

71. Theodore C. Henry, *Addresses on "Kansas Stock Interests" and "Kansas Forestry"* (Abilene, KS: Gazette Steam Printing Office, 1882), 4 (Kansas State Historical Society). The transcript of Henry's speeches misspells the governor's name as "Tarases."

72. Henry, "Addresses," 4 for climate, 2 for "physical phenomena.

73. Dorsey, "Land Stealing," 405. Dorsey wrote, "Nothing is more idle than the talk that can be heard on all sides respecting the rain-fall increasing within what is known as the arid region."

74. For climate change and contemporary capitalism, see Naomi Klein, *This Changes Everything: Capitalism vs. the Climate* (New York: Simon and Schuster, 2014).

75. Vladimir Jankovic and Andrew Bowman, "After the Green Gold Rush: The Construction of Climate Change as a Market Transition," *Economy and Society* 43, no. 2 (2014): 233–59.

Chapter 3

1. Francis E. Nipher, "Report on Missouri Rainfall, with Averages for Ten Years Ending December 1887," *Transactions of the Academy of Science of St. Louis*, vol. 5, *1886–1891* (St. Louis, MO: R. P. Studley, 1892), 383.

2. Some nineteenth-century Americans used the term "middle border" to describe prairie states as well as much of the West. Author Hamlin Garland popularized the term in the late nineteenth and early twentieth centuries. See Hamlin Garland, *A Son of the Middle Border* (New York: Macmillan, 1917).

3. James Rodger Fleming, *Meteorology in America, 1800–1870* (Baltimore: Johns Hopkins University Press, 1990), xxi–xxii; see also Fleming, *Historical Perspectives on Climate Change* (Oxford: Oxford University Press, 1998), 33–41.

4. See Zeke Baker, "Agricultural Capitalism, Climatology and the 'Stabilization' of Climate in the United States, 1850–1920," *British Journal of Sociology*, June 2020, 1–18.

5. Paul Edwards, "Meteorology as Infrastructural Globalism," *Osiris* 21 (2006): 230.

6. For "pure science" in the Gilded Age, see Paul Lucier, "The Professional and the Scientist in Nineteenth-Century America," *Isis* 100 (2009): 723. See also Lucier, "The Origins of Pure and Applied Science in Gilded Age America," *Isis* 103 (2012): 527–36.

7. Philipp Lehmann, "Whither Climatology? Brückner's *Climate Oscillations*, Data Debates, and Dynamic Climatology," *History of Meteorology* 15 (2015): 51.

8. In his study of provincial meteorology in Cornwall, Naylor has argued that mass-scale collection of meteorological data gave rise to "very real concerns in the late 1860s about how to turn continuous records into numerical results useful to science and government." See Simon Naylor, "Nationalizing Provincializing Weather: Meteorology in Nineteenth-Century Cornwall," *British Journal for the History of Science* 39 (2006): 419.

9. I consider the notion of "everyday science" an amalgam of the concepts of "vernacular science," "citizen science," "democratic science," and "popular science." My aim in using these expansive and admittedly nebulous terms is to highlight the unevenness of late nineteenth-century scientific bureaucratization, professionalization, systematization, centralization, and standardization. As Kathleen Pandora has argued, "vernacular discursive forms" of science served as a kind of "intellectual commons" where "social and theoretical comment can circulate without regard for scientific property." See Kath-

erine Pandora, "Knowledge Held in Common: Tales of Luther Burbank and Science in the American Vernacular," *Isis* 92 (September 2001): 492. For institutional climate science, see Fleming, *Meteorology in America*. For antebellum-era "natural inquiry" and settler perceptions, see Conevery Bolton Valenčius, *The Health of the Country: How American Settlers Understood Themselves and Their Land* (New York: Basic Books, 2002); Valenčius, *The Lost History of the New Madrid Earthquakes* (Chicago: University of Chicago Press, 2013), especially 10, 17, and 177. In her study on folklore and meteorological knowledge in Switzerland, Sarah Strauss argues that "folklore, as well as individual practice, is embedded into the scientific process." See Strauss, "Weather Wise: Speaking Folklore to Science in Leukerbad," in *Weather, Climate, Culture*, ed. Sarah Strauss and Benjamin Orlove (New York: Berg, 2003), 52–53.

10. See Kristine C. Harper, *Make It Rain: State Control of the Atmosphere in Twentieth-Century America* (Chicago: University of Chicago Press, 2017), 15. Harper explains how, in the late 1800s and early 1900s, agencies such as the Weather Bureau "had very little credibility," partly because there were still no formally trained "expert" meteorologists. In *Victorian Popularizers of Knowledge: Designing Nature for New Audiences* (Chicago: University of Chicago Press, 2007), Bernard Lightman analyzes the relation between popular science and elite science: "Popular culture can actively produce its own indigenous science, or can transform the products of elite culture in the process of appropriating them" (14).

11. For "men of science" deriving inspiration from folklore and anecdote, see John Hay, "Atmospheric Absorption and Its Effect upon Agriculture," *Proceedings of the Eighteenth Annual Meeting 1890 of the Kansas State Board of Agriculture* (Topeka, KS, 1890); Harvey Culbertson, "Meteorology," *Annual Report of the Nebraska State Horticultural Society 1885* (Lincoln: State Journal Company, 1887). By using "professional" in this context, I am using the current meaning of the term to refer to meteorological and climatic thinkers who earned salaries for their scientific endeavors. For a study on the complex meaning of "professional" in the Gilded Age, see Lucier, "Professional and the Scientist." Lucier argues that late nineteenth-century "men of science" sought to distinguish themselves from "professionals" tainted by economic dealings.

12. For Hinrichs's defense of "real science," see Hinrichs, "Faith and Science," lecture delivered by Gustavus Hinrichs before the students of the [Iowa] State University, undated (likely 1867), G. Hinrichs Papers, box 1, University of Iowa Library, University Archives, RG 99.0039. For Hinrichs's skepticism, see Hinrichs, "Rainfall and Timber in Iowa," *Transactions of the Iowa State Horticultural Society for 1879* (Des Moines, IA: F. M. Mills, 1880), 199–200.

13. See Stephen F. Corfidi, Michael C. Coniglio, Ariel E. Cohen, and Corey M. Mead, "A Proposed Revision to the Definition of 'Derecho,'" *Bulletin of the American Meteorological Society* 97, no. 6 (June 2016): 935–49; W. P. Palmer, "Dissent at the University of Iowa: Gustavus Detlef Hinrichs—Chemist and Polymath," *Chemistry* 16, no. 6 (2007): 534–53.

14. For biographical information on Hinrichs, see the obituary by Charles Keyes in *Iowa Academy of Science* 30, (1923): 28–31, G. Hinrichs Papers, box 1, biography folder, University of Iowa Library, University Archives, RG 99.0039.

15. Gustavus Hinrichs, *Rainfall Laws Deduced from Twenty Years of Observation* (Washington, DC: Weather Bureau, 1893), 77.

16. Hinrichs moved to St. Louis in 1889 and began teaching at Washington University. In his *Sixth Biennial Report of the Central Station of the Iowa Weather Service* (Des

Moines, IA: G. H. Ragsdale, 1889), Hinrichs criticized the Weather Bureau for its poor scientific practices and its efforts to undermine his project. In one of his first reports, J. R. Sage, head of the Weather Bureau-affiliated Iowa Weather and Crop Service, fired back against Hinrichs's organization, citing the "defect in its management" and pointing out that the Iowa General Assembly had shifted its support from Hinrichs to his rivals. *Annual Report of the Iowa Weather and Crop Service in Co-operation with the U.S. Department of Agriculture, Weather Bureau, for the Meteorological Year 1890* (Des Moines, IA: G. H. Ragsdale, 1891), 6–7.

17. Gustavus Hinrichs, "A Few Facts about the Iowa Weather Service," February 2, 1888, Gustavus Hinrichs Papers, box 3, University of Iowa Library, University Archives, RG 99.0039. Although Hinrichs's institutional home, the State University of Iowa (later known as the University of Iowa), predated the Morrill Act of 1862, his utilitarian outlook fits the guiding ethos of the new land grant schools. Hinrichs's emphasis on usable knowledge also mirrors the approach of late nineteenth- and early twentieth-century "field stations" chronicled by Jeremy Vetter in *Field Life: Science in the American West during the Railroad Era* (Pittsburgh, PA: University of Pittsburgh Press, 2016).

18. Gustavus Hinrichs, *Second Annual Report of the Iowa State Weather Service*, appendix to the *Report of the Iowa State Agricultural Society* for the year 1877, 624.

19. Hinrichs, "Rainfall and Timber in Iowa," 199–200.

20. For "everyday science," see Valenčius, "Lost History of the New Madrid Earthquakes," 10. For a discussion of "pure science," see Lucier, "Origins of Pure and Applied Science."

21. Hinrichs, *Rainfall Laws* 17 for "large garden," 13 for "thrashing machine," and 15 for "total utilizable."

22. See George E. Curtis, "Review of *Rainfall Laws, Deduced from Twenty Years of Observation*," *American Meteorological Journal* 10 (April 1894). For a study on the politics of probability and prediction in turn-of-the-century American meteorology, see Jamie L. Pietruska, "US Weather Bureau Chief Willis Moore and the Reimagination of Uncertainty in Long-Range Forecasting," *Environment and History* 17, no. 1 (February 2011): 79–105.

23. See Gustavus Hinrichs, *Fifth Biennial Report of the Central Station of the Iowa Weather Service* (Des Moines, IA: Geo. E. Roberts, 1887), 5.

24. Jonathan Levy, *Freaks of Fortune: The Emerging World of Capitalism and Risk in America* (Cambridge, MA: Harvard University Press, 2012), especially 79–82; Jamie L. Pietruska, *Looking Forward: Prediction and Uncertainty in Modern America* (Chicago: University of Chicago Press, 2017), 71–107. For probability, luck, and chance, see also Ian Hacking, *The Taming of Chance* (Cambridge: Cambridge University Press, 1990), and T. J. Jackson Lears, *Something for Nothing: Luck in America* (New York: Viking, 2003).

25. See *Daily Register*, July 6, 1875, Hinrichs Papers, box 1, and "Documents Relating to the Dismissal of Dr. Gustavus Hinrichs," Hinrichs Papers, box 1.

26. "Documents Relating to . . . ," Hinrichs Papers, box 1.

27. Keyes, obituary, *Iowa Academy of Science* 30 (1923).

28. James Bergman, "Knowing Their Place: The Blue Hill Observatory and the Value of Local Knowledge in an Era of Synoptic Weather Forecasting, 1884–1894," *Science in Context* 29, no. 3 (September 2016): 305–46.

29. Phaedra Daipha, "Weathering Risk: Uncertainty, Weather Forecasting, and Expertise," *Sociology Compass* 6 (2012): 18. For scale and climatology, see also Deborah Coen, "Imperial Climatographies from Tyrol to Turkestan," *Osiris* 26 (2011): 45–65.

30. Marlene Bradford, "Historical Roots of Modern Tornado Forecasts and Warnings," *Weather and Forecasting* 14 (August 1999): 484–91; Peter J. Thuesen, *Tornado God: American Religion and Violent Weather* (New York: Oxford University Press, 2020), 69–100.

31. For "destructive great storms," see Hinrichs, *Second Annual Report*, 655. See also Hinrichs, "Tornadoes and Derechos," *American Meteorological Journal* 5, no. 9 (January 1889).

32. Gustavus Hinrichs, *First Biennial Report of the Central Station of the Iowa Weather Service* (Des Moines, IA: F. M. Mills, 1880), 22–23. For an analysis of "field networks" and the influence of telegraphs on weather observation, see Jeremy Vetter, "Lay Observers, Telegraph Lines, and Kansas Weather: The Field Network as a Mode of Knowledge Production," *Science in Context* 24, no. 2 (June 2011): 259–80.

33. Hinrichs, *Second Annual Report*, 623.

34. For "simplify and systematize," see *First Annual Report*, 22. For "modern science" see Gustavus Hinrichs, *American Scientific Monthly* (edited and published by Prof. Gustavus Hinrichs), 1, no. 1 (July 1870), State Historical Society of Iowa (Q1.A8), 3.

35. David Cahan describes the solidifying of scientific titles and designations in nineteenth-century science in "Looking at Nineteenth-Century Science: An Introduction," in *From Natural Philosophy to the Sciences: Writing the History of Nineteenth-Century Science*, ed. David Cahan (Chicago: University of Chicago Press, 2003), 4; see also "Institutions and Communities," in Cahan, *From Natural Philosophy to the Sciences*, 297. Along with "full-time devotion to and pay for scientific work" and "advanced well-defined educational credentials," Cahan views the university as the "principal institutional setting for science." It is noteworthy that even though Hinrichs worked for the State University of Iowa, his weather service remained somewhat independent from the university's institutional umbrella.

36. For an analysis of center-periphery tensions in meteorological science, see Jamie L. Pietruska, "Hurricanes, Crops, Capital: The Meteorological Infrastructure of American Empire in the West Indies," *Journal of the Gilded Age and Progressive Era* 15 (October 2016): 410.

37. See Ralph R. Hamerla, *An American Scientist on the Research Frontier: Edward Morley, Community, and Radical Ideas in Nineteenth-Century Science* (Dordrecht, Netherlands: Springer, 2006), 2; Naylor, "Nationalizing Provincial Science," 409.

38. The concept of "center of calculation" is from Bruno Latour, *Science in Action: How to Follow Scientists and Engineers through Society* (Cambridge, MA: Harvard University Press, 1987).

39. Hinrichs's home was built in 1879. See handwritten note on an 1909 photograph of Hinrichs's house in University of Iowa archives, Hinrichs Papers, photographs file. Before his house was built Hinrichs used another observatory at Church and Clinton Streets in Iowa City. See Ray Wolf's "Brief History of Gustavus Hinrichs, Discoverer of the Derecho," http://www.spc.noaa.gov/misc/AbtDerechos/hinrichs/hinrichs.htm (accessed January 19, 2021).

40. David Cahan discusses the formation of "imagined communities of science" in the nineteenth century. See David Cahan, "Introduction," in *From Natural Philosophy to the Sciences*, 11.

41. "Neglect" is from Hinrichs, *First Biennial Report*, 6. "Estrange" is from Hinrichs, *Sixth Biennial Report*, 5. "Clerical help" is from Hinrichs, *First Biennial Report*, 24.

42. Hinrichs, *Second Biennial Report of the Central Station of the Iowa Weather Service* (Des Moines: F. M. Mills, 1882), 31.

43. Hinrichs, *Second Annual Report*, 622.

44. Gustavus Hinrichs, *Notes on Cloud Forms and the Climate of Iowa*, Central Station, special bulletin (Iowa City, IA: Iowa Weather Service, 1883).

45. Hinrichs, *Second Annual Report*, 625. Although the state of Iowa provided some financial support after 1878, Hinrichs shouldered much of the weather service's financial burden himself. See Wolf, "Brief History of Gustavus Hinrichs."

46. Hinrichs, *Third Biennial Report of the Central Station of the Iowa Weather Service* (Des Moines, IA: George E. Roberts, 1883), 5.

47. For an overview of the dramatic rise of popular science in the Gilded Age United States, see Rebecca Edwards, *New Spirits: Americans in the Gilded Age, 1865–1905* (Oxford: Oxford University Press, 2006), 151–69. Citing magazines such as *Popular Science Monthly*, *Scientific American*, and *Science*, Edwards argues that "America became a nation of scientific enthusiasts" (160).

48. Hinrichs, *Third Biennial Report*, 5. Hinrichs's doubt over his work's significance underscores historian Jeremy Vetter's argument about the paradoxical nature of scientific universalism. "The desire to attain knowledge that can be applied to all times and places," he writes, "has been an overriding ambition of modern science." But the drive to create universalist knowledge "has not always worked to produce pragmatically useful and environmentally sustainable knowledge on the ground in particular places." See Vetter, *Field Life*, 338.

49. C[orydon] P. Cronk, "Influence of Forests on Climate and Agriculture," *Maryland State Weather Service Monthly Report* 3, no. 6 (Washington, DC: US Department of Agriculture, Weather Bureau, October 1893), 57–58.

50. Hinrichs, *Second Annual Report*, 624.

51. Paul Travis, "Changing Climate in Kansas: A Late 19th-Century Myth," *Kansas History* 1, no. 1 (Spring 1978): 50.

52. Charles Withers and David Livingstone, "Thinking Geographically about Nineteenth-Century Science," in *Geographies of Nineteenth-Century Science*, ed. Withers and Livingstone (Chicago: University of Chicago Press, 2011), 5.

53. Cronk, "Influence of Forests," 58. Cronk may be referring to the federal Timber Culture Act of 1873 or to other afforestation initiatives such as the Arbor Day movement.

54. See Thomas Gieryn, *Cultural Boundaries of Science: Credibility on the Line* (Chicago: University of Chicago Press, 1999), 15–18. Gieryn identifies three types of boundary work. My main concern here is with the category he terms "exclusion"—the exclusion of certain figures, methods, and practices as unscientific.

55. For a discussion of scientific "translation," see Helen Tilley, "Global Histories, African Genealogies, and Vernacular Science, or Is the History of Science Ready for the World?" *Isis* 101, no. 1 (March 2010): 112, 117. As Sara B. Pritchard has argued, boundary work "has the potential to be generative and empowering, sometimes even counter-hegemonic." See Pritchard, "Joining Environmental History with Science and Technology Studies: Promises, Challenges, and Contributions," in *New Natures*, ed. Sara B. Pritchard, Finn Arne Jorgensen, and Dolly Jorgensen (Pittsburgh, PA: University of Pittsburgh Press, 2013), 13.

56. Letter from S. L. Dosher, *American Meteorological Journal* 9, no. 10 (February 1893).

57. Isaac P. Noyes, "A New View of the Weather Question," *Kansas City Review of Science and Industry*, vol. 2 (Kansas City, MO: Ramsey, Millet, and Hudson, 1885); see 218–19 for "weather mystery," 227 for a discussion of climate change.

58. Hinrichs, *Rainfall Laws*; for another 1890s source using measurements to endorse climatic influences, see L. C. Corbett, "Influence of Groves on the Moisture Content of the Air," *Forester* (West Virginia University) 3, no. 4 (April 1, 1897).

59. For a description of forest-climate influence theories as "moonshine notions," see J. W. Foster's remarks in "Discussion of Meteorology," *Transactions of the Iowa State Horticultural Society for 1879* (Des Moines, IA: F. M. Mills, 1880), 486.

60. *Topeka (KS) Weekly Tribune*, March 27, 1879, Kansas State Historical Society, "Rain and Rainfall" Clippings.

61. See, for example, remarks by "Mr. Holton" quoted in Charles Dana Wilber, *Great Valleys and Prairies of Nebraska and the Northwest* (Omaha, NE: Daily Republican Print, 1881), 81. Holton voiced his disdain for "philosophers" and "a certain class of scientists," experts who disputed theories of human-induced climatic improvement.

62. John Trowbridge, "Great Fires and Rain-Storms," *Popular Science Monthly*, December 1872, 206, 211. The Great Chicago Fire, Peshtigo fire, and other massive conflagrations of the early 1870s spurred an interest in investigating the role of fires in shaping weather patterns. See also Jonathan Periam, "Forest Tree Planting, as a Means of Wealth," *Transactions of the Illinois State Horticultural Society for 1871* (Chicago: Reade, Brewster, 1872), 34–35. For a discussion of older theories on fire and climate, see James Rodger Fleming, *Fixing the Sky: The Checkered History of Weather and Climate* (New York: Columbia University Press, 2010), 56–59.

63. Historian of science Ronald Kline has highlighted the fluid, ever-evolving meaning of categories such as "applied science." See Ronald Kline, "Construing 'Technology' as 'Applied Science': Public Rhetoric of Scientists and Engineers in the United States, 1880–1945," *Isis* 86, no. 2 (June 1995): 194–98.

64. J. L. Budd [Iowa Board of Forestry], "Possible Modification of Our Prairie Climate," *Sixth Annual Meeting of the American Forestry Congress* (Springfield, IL: State Register Book and Job Print, 1887), 20–22.

65. *Larned (KS) Press*, August 8, 1878.

66. For biographical information on Snow, see F. H. Snow file 2/6/6 and finding aid, Spencer Library, University of Kansas.

67. F. H. Snow, "Climate of Kansas," report submitted to Alfred Gray, Secretary State Agricultural Society, January 1st, 1873; see 1 for "instruments" and "apparatus" and 8 for "oldest residents." Snow alluded only to the memories of the oldest Euro-American residents, entirely eliding Native American experience and knowledge.

68. Snow, "Climate of Kansas," 7–8.

69. A letter from F. H. Snow is cited in Clinton Carter Hutchinson's *Resources of Kansas: Fifteen Years Experience* (Topeka, KS: author, 1871), 37.

70. F. H. Snow, "Is the Rainfall of Kansas Increasing?" *Kansas City Review of Science and Industry*, vol. 8 (Kansas City, MO: Ramsey, Millet, and Hudson, 1885), 458.

71. For a discussion of "regimes of high uncertainty" in meteorology, see Daipha, "Weathering Risk," 15–16.

72. F. H. Snow, "Periodicity of Kansas Rainfall and Possibilities of Storage of Excess Rainfall," *Ninth Biennial Report of the Kansas State Board of Agriculture* (Topeka: Edwin H. Snow, 1895), 338–39. For another example of a Kansan espousing theories of seven-year climatic cycles, see Edward Charles Murphy, "Is the Rainfall in Kansas Increasing?" *Transactions of the Kansas Academy of Science*, vol. 13 (Topeka, KS: Hamilton Printing Company, 1893), 19. See Lehmann, "Whither Climatology?" for a study of repeating patterns and oscillations in turn-of-the-century climatology.

73. Snow, "Periodicity," 339–40.

74. For a discussion of geographies of health in the late nineteenth-century West, see Gregg Mitman, "Geographies of Hope: Mining the Frontiers of Health in Denver and Beyond, 1870–1965," *Osiris* 19 (2004): 93–111.

75. M. V. B. Knox, "Climate and Brains," *Transactions of the Kansas Academy of Science*, vol. 5 (Topeka, KS: George W. Martin, 1877), 5–9.

76. Hinrichs, *First Biennial Report*, 5.

77. J. W. Foster, *The Mississippi Valley* (Chicago: S. C. Griggs, 1869), xi; see also 356–57.

78. David Singerman, "Science, Commodities, and Corruption in the Gilded Age," *Journal of the Gilded Age and Progressive Era* 15, no. 3 (2016): 290.

Chapter 4

1. F[loyd] P[erry] Baker, *Preliminary Report on the Forestry of the Mississippi Valley and Tree Planting on the Plains* (Washington, DC: GPO, 1883), 22.

2. J. T. Allan, *Nebraska and Its Settlers: What They Have Done and How They Do It; Its Crops and People* (Omaha, NE: Union Pacific Company Land Department, 1883), 8.

3. Char Miller, "French Lessons: F. P. Baker, American Forestry, and the 1878 Paris Universal Exposition," *Forest History Today*, Spring/Fall 2005, 10–11.

4. See F. B. Hough, *Report upon Forestry*, vol. 1 (Washington, DC: GPO, 1878), 48.

5. Theodore C. Henry, *Addresses on "Kansas Stock Interests" and "Kansas Forestry"* (Abilene, KS: Gazette Steam Printing Office, 1882), 2. Known as the Kansas wheat king, Henry spent much of the 1870s experimenting with new harvesters and other machines on his extensive farms in Kansas. Frustrated, he moved to the Front Range of Colorado and took up the irrigation ditch business. See the *Wichita Eagle*, January 18, 1901, and Floyd Benjamin Streeter, *The Kaw: The Heart of a Nation* (New York: Farrar and Rinehart, 1941), 237–39.

6. *Topeka Commonwealth*, February 7, 1882.

7. For an example of climate maps derived from Signal Corps data, see Elias Loomis, "Contributions to Meteorology, Being Results Derived from an Examination of the Observations of the United States Signal Corps, and from Other Sources," *American Journal of Science and Arts*, 3rd ser., 13, no. 73 (January 1877). For a later study also based on government data, see Adolphus Greely, "Report on the Climatology of the Arid Region of the United States, with Reference to Irrigation," in *Report on the Climate of New Mexico* by the Chief Signal Officer of the Signal Corps (Washington, DC: GPO, 1891).

8. Donald Pisani, *Water, Land, and Law in the West: The Limits of Public Policy, 1850–1920* (Lawrence: University of Kansas Press, 1996), 132.

9. For maps as "slippery" and for a discussion of appropriation in cartography, see, among others, J. B. Harley, "Texts and Contexts in the Interpretation of Early Maps," in *The New Nature of Maps: Essays in the History of Cartography*, ed. Paul Laxton (Baltimore: Johns Hopkins University Press, 2001), 34–38; Denis Wood, *The Power of Maps* (New York: Guilford Press, 1992), 25; Bülent Batuman, "The Shape of the Nation: Visual Production of Nationalism through Maps in Turkey," *Political Geography* 29, no. 4 (2010): 230.

10. See Susan Schulten, *Mapping the Nation: History and Cartography in Nineteenth-Century America* (Chicago: University of Chicago Press, 2012). Isopleth mapping, pioneered by Alexander von Humboldt in the early nineteenth century, uses isolines

to bound areas that fall within a certain data range for the salient statistical variable. Isopleth maps showing rainfall are known as isohyetal maps, and the lines used to denote areas with equal rainfall are termed isohyets. Choropleth mapping, on the other hand, uses predetermined statistical areas such as counties or states and colors them according to data ranges. For discussions on the origins of isopleth mapping, see Schulten, *Mapping the Nation*, 80–86; Katharine Anderson, "Mapping Meteorology," in *Intimate Universality: Local and Global Themes in the History of Weather and Climate*, ed. James Rodger Fleming, Vladimir Jankovic, and Deborah R. Coen (Sagamore Beach, MA: Science History Publications, 2006), 70–71; Aaron Sachs, *The Humboldt Current: Nineteenth-Century Exploration and the Roots of American Environmentalism* (New York: Viking, 2006), 13, 51.

11. Anderson, "Mapping Meteorology," 70–71.

12. Schulten, *Mapping the Nation*, 4, 7. See also Gregory H. Nobles, "Straight Lines and Stability: Mapping the Political Order of the Anglo-American Frontier," *Journal of American History* 80, no. 1 (June 1993): 9–35; David Bernstein, *How the West Was Drawn: Mapping, Indians, and the Construction of the Trans-Mississippi West* (Lincoln: University of Nebraska Press, 2018). Benedict Anderson discusses the role of maps in nationalist projects in *Imagined Communities: Reflections on the Origin and Spread of Nationalism*, rev. ed. (London: Verso, 2006); see especially 163–86.

13. See Raymond Craib, *Cartographic Mexico: A History of State Fixations and Fugitive Landscapes* (Durham, NC: Duke University Press, 2004); Valerie Kivelson, *Cartographies of Tsardom: The Land and Its Meanings in Seventeenth-Century Russia* (Ithaca, NY: Cornell University Press, 2006); Denis Wood, "Mapmaking, Counter-Mapping, and Map Art in the Mapping of Palestine," in *Rethinking the Power of Maps*, ed. Denis Wood, John Fels, and John Krygier (New York: Guilford Press, 2010); Nancy Lee Peluso, "Whose Woods Are These? Counter-Mapping Forest Territories in Kalimantan, Indonesia," in *The Anthropology of Development and Globalization*, ed. Marc Edelman and Angelique Haugerud (Malden, MA: Blackwell, 2005).

14. Wood, *Power of Maps*, 22.

15. As James D. Drake shows in his study of geographic and continental perceptions in colonial and early republic America, "voids in geographic knowledge" can sometimes empower cartographers. See Drake, *The Nation's Nature: How Continental Presumptions Gave Rise to the United States of America* (Charlottesville: University of Virginia Press, 2011), 28–29.

16. Martin Brückner, *The Geographic Revolution in Early America: Maps, Literacy, and National Identity* (Chapel Hill: University of North Carolina Press, 2006), 238–39, 12–13.

17. In her book on the history of farming on the northern Great Plains, Mary Hargreaves explains that the desert loomed large in the mental maps of many Americans: "Until the 1850s, school geographies and atlases repeated the legend of the 'Great Desert' applied by Major Stephen Harriman Long to the district between the Platte and the Canadian Rivers." Hargreaves argues that the belief in the Great American Desert went "virtually unchallenged" as Santa Fe traders, Mormons, and Gold Rushers all attested to the "hardships of the plains." Mary Wilma Hargreaves, *Dry Farming in the Northern Great Plains, 1900–1925* (Cambridge, MA: Harvard University Press, 1957), 26. See also Martyn J. Bowden, "The Great American Desert in the American Mind: The Historiography of a Geographical Notion," in *Geographies of the Mind: Essays in Historical Geosophy*, ed. David Lowenthal and Martyn Bowden (New York: Oxford University Press, 1976); Elliott West, *The Contested Plains: Indians, Goldseekers and the Rush to Colorado*

(Lawrence: University Press of Kansas, 1998); B. H. Baltensperger, "Plains Boomers and the Creation of the Great American Desert Myth," *Journal of Historical Geography* 18, no. 1 (1992): 59–73. For a Gilded Age discussion of the origins of the "Great American Desert" myth, see Clinton Carter Hutchinson, *Resources of Kansas: Fifteen Years Experience* (Topeka, KS: author, 1871).

18. Sara T. L. Robinson, *Kansas: Its Interior and Exterior Life, Including a Full View of Its Settlement, Political History, Social Life, Climate, Soil, Productions, Scenery Etc.* (Boston: Crosby, Nichols, 1856), 2.

19. Perkins is quoted at length in L. P. Brockett, *Our Western Empire, or The New West Beyond the Mississippi* (Columbus, OH: William Garretson, 1881), 39–41. The "drouth line" quotation is from "One Hundredth Meridian" (pamphlet), *Topeka (KS) Commonwealth* (1879), Kansas Historical Society Pamphlet Collection.

20. *Topeka (KS) Commonwealth*. For more on isothermal lines and slavery, see David Zarefsky, *Lincoln, Douglas, and Slavery* (Chicago: University of Chicago Press, 1993), 91. For another example of an author's railing against the "Great American Desert," see Patrick Hamilton, *The Resources of Arizona* (San Francisco, CA: A. L. Bancroft, 1883). Hamilton denounces the "wiseacres . . . who included all that vast and fertile domain west of the Missouri in the "Great American Desert" (191). Hamilton's discussion of Arizona reveals that questions about the "Great American Desert" and climate change extended beyond the Great Plains.

21. Brockett, *Our Western Empire*, 41.

22. Cyrus Thomas in F. V. Hayden, *Sixth Annual Report of the United States Geological Survey of the Territories* (Washington, DC: GPO, 1873), 278.

23. Hugh Rankin Hilton, "Effects of Civilization on the Climate and Rain Supply of Kansas," Lecture Delivered by H. R. Hilton, Esq., before the Scientific Club of Topeka, Wednesday Evening, March 31st, 1880, Spencer Library, University of Kansas Archives, RH C4318, 2–3.

24. For "imaginary line," see Martin Allen, "Tree Planting on the Plains," in *American Journal of Forestry* (Cincinnati, OH: Robert Clarke, 1882–83), 1:299. For "varying and uncertain," see Martin Allen, "Brief Historical Sketch of Ellis County up to the Close of the Centennial Year," *Hays City Sentinel*, January 18, 1878; n.b., this Allen is not to be confused with J. T. Allan, a writer mentioned earlier.

25. For a discussion of twentieth-century efforts to map change over time as well as risk, see Mark Monmonier, *Cartographies of Danger: Mapping Hazards in America* (Chicago: University of Chicago Press, 1997), especially 65–87. See also Schulten, *Mapping the Nation*, 23 and 72.

26. For biographical information on Wilber, see E. A. Kral, "Charles Dana Wilber: Scientific Promoter, Pioneer of the West and Town Founder," *Wilber Republican*, August 2, 2000.

27. As the figure who coined and popularized the phrase "rain follows the plow," Wilber has received cursory mentions in nearly every source that touches on Gilded Age climate theory. The most thorough treatment of Wilber's writings is probably in David Emmons, *Garden in the Grasslands: Boomer Literature of the Central Great Plains* (Lincoln: University of Nebraska Press, 1971), 138–39, 186.

28. Charles Dana Wilber, *The Great Valleys and Prairies of Nebraska and the Northwest* (Omaha, NE: Daily Republican Print, 1881), 52.

29. For Hinrichs's belief in the fundamental uncertainty of science, see, for example, Gustavus Hinrichs, *Rainfall Laws Deduced from Twenty Years of Observation* (Washington, DC: Weather Bureau, 1893), 11 and 82.

30. Hinrichs, *Second Annual Report*, 624. See also Gustavus Hinrichs, "Rainfall and Timber in Iowa," *Transactions of the Iowa State Horticultural Society for 1879* (Des Moines, IA: F. M. Mills, 1880), 199.

31. Hinrichs, *Second Annual Report*, 624.

32. Wood, *Power of Maps*, 25.

33. James Rodger Fleming, *Meteorology in America, 1800–1870* (Baltimore: Johns Hopkins University Press, 1990), 129.

34. Charles A. Schott, "Rain Chart of the United States Showing by Isohyetal Lines the Distribution of the Mean Annual Precipitation in Rain and Melted Snow," 1868. I consulted a later copy of the original map. See "Rain Chart of the United States. . . . This is a copy of the chart constructed for the Smithsonian Institution in 1868 by Charles A. Schott except that the lines are slightly modified in Texas and New Mexico," Department of the Interior, US Geographical and Geological Survey of the Rocky Mountain Region, J. W. Powell in charge.

35. "Rainfall-Chart of the United States Showing the Distribution by Isohyetal Curves of the Mean Precipitation in Rain and Melted Snow for the Year, mean annual precipitation shown by isohyetal curves for every sixth inch from 8 to 68 inches," 2nd ed., Smithsonian Institution, Prof. J. Henry, secretary, including records to 1877. Prof. Spencer F. Baird, Secretary. Constructed by Charles A. Schott, Assistant USC and G Survey (Washington, DC, 1880).

36. Henry Allen Hazen, "Variation of Rainfall West of the Mississippi River," Signal Service Notes 7 (Washington, DC: War Department, Office of the Chief Signal Officer, 1883), 4.

37. For an analysis of the power of cartographic voids and unknowns, see Barbara Belyea, "Inland Journeys, Native Maps," in *Cartographic Encounters: Perspectives on Native American Mapmaking and Map Use*, ed. Malcolm G. Lewis (Chicago: University of Chicago Press, 1998).

38. "Disgruntled" is from Fleming, *Meteorology in America*, 110–11. Blodget worked at the Smithsonian intermittently in the 1850s and also worked at the War Department.

39. The first quotation is from Lorin Blodget, *Climatology of the United States* (Philadelphia: J. B. Lippincott, 1857), 85. The other two are from Blodget, "Forest Cultivation on the Plains," *Report of the Commissioner of Agriculture for the Year 1872* (Washington, DC: GPO, 1872), 318 and 325.

40. Mark Harrington, "Is the Rain-Fall Increasing on the Plains?" *American Meteorological Journal* 4, no. 8 (December 1887): 370–73.

41. "The Influence of Forests on the Quantity and Frequency of Rainfall," *Science* 12, no. 303 (November 23, 1888): 244.

42. Henry Gannett, "Do Forests Influence Rainfall?," *Science* 11, no. 257 (January 6, 1888): 3.

43. Gannett, "Influence of Forests," 243.

44. J. D. Whitney, "Plain, Prairie, Forest," *American Naturalist* 10, no. 10 (October 1876): 579; J. D. Whitney, "Plain, Prairie, Forest," *American Naturalist* 10, no. 11 (November 1876): 656.

45. Whitney, "Plain, Prairie, Forest" (October), 579–80.

46. Greely referred to climate change as a "vexed" question in his book *American Weather* (New York: Dodd, Mead, 1888), 151.

47. John P. Finley, *Certain Climatic Features of the Two Dakotas* (Washington, DC: GPO, 1893), 8–12.

48. *Report of Rainfall in Washington Territory, Oregon, California, Idaho, Nevada, Utah, Arizona, Colorado, Wyoming, New Mexico, Indian Territory, and Texas, for from Two to Forty Years*, H.R. 50th Cong., 1st Sess., Ex. Doc. No. 91 (Washington, DC: GPO, 1889), 6, 12.

49. For "make sense of the natural world," see Schulten, *Mapping the Nation*, 107.

50. The "mirror of nature" is from Harley, "Deconstructing the Map," 4.

Interlude

1. Robert Dyrenforth, "Can We Make It Rain?" *North American Review* 153, no. 419 (October 1891): the list of supplies is on 394. Dyrenforth's piece in the *North American Review* includes a lengthy narrative describing the experiment. Most of the supplies were purchased with a $9,000 allocation from Congress secured for the experiment by Dyrenforth's ally Charles B. Farwell, a senator from Illinois. Large ranchers and other supportive Texas businessmen also pitched in, donating some of the explosives. See Michael R. Whitaker, "Making War on Jupiter Pluvius: The Culture and Science of Rainmaking in the Southern Great Plains, 1870–1913," *Great Plains Quarterly* 33, no. 4 (Fall 2013): 209.

2. Dyrenforth was a Washington DC–based patent lawyer and Civil War veteran. Although many sources refer to him as "General" Dyrenforth, it seems he never ascended above the rank of major in the Union Army. See James Rodger Fleming, *Fixing the Sky: The Long and Checkered History of Weather and Climate Control* (New York: Columbia University Press, 2010), 65; Whitaker, "Making War," 217n9.

3. Whitaker, "Making War," 209.

4. See Julie Courtwright, "On the Edge of the Possible: Artificial Rainmaking and the Extension of Hope on the Great Plains," *Agricultural History* 89, no. 4 (Fall 2015): 536–58. Courtwright argues that late nineteenth-century Americans enjoyed entertainment that invited the temporary suspension of disbelief as well as introspection about the nature of truth and falsehood. For further analysis of early rainmaking attempts, see Fleming, *Fixing the Sky*, and Kristine C. Harper, *Make It Rain: State Control of the Atmosphere in Twentieth-Century America* (Chicago: University of Chicago Press, 2017), 21–25.

5. Dyrenforth, "Can We Make It Rain?," 393.

6. The *Phillips County (KS) Inter Ocean* (August 1891, KSHS–"Rain and Rainfall" Clippings, 551.57R) offered an overview of Curtis and Dyrenforth's scientific rationales for the experiment. Dyrenforth explained how percussion and other human influences modified climate in "Can We Make It Rain?," 387.

7. Dyrenforth, "Can We Make It Rain?," 393–94 for the battle lines and 397 for Dyrenforth's strategy of coaxing rain from pregnant clouds. Whitaker depicts percussive rainmaking experiments like Dyrenforth's as motivated by the collective experience of Civil War veterans. See Whitaker, "Making War," 208–9.

8. Part of Dyrenforth's approach involved simulating wartime conditions as accurately as possible. Influenced by the writings of Edward Powers, Dyrenforth believed that artillery exchanges during battles had caused storms and rainfall. He cited myriad examples of postbattle rainfall from the Napoleonic Wars, the US Civil War, and the Franco-Prussian War. See Dyrenforth, "Can We Make It Rain?," 388–89. Powers's widely read *War and the Weather* outlined the percussive artillery simulation approach that Dyrenforth would use. See Edward Powers, *War and the Weather* (1871; Delavan, WI:

E. Powers, 1890), 100–101. Powers's book emphasized the testimony of soldiers more than abstract theories explaining the origin of rainfall; his scientific approach emphasized experience and experiment over theory: "Facts are of more importance than theories, but in order that a knowledge of facts may lead to the most useful results it is necessary that they should be supplemented by reasonable theories." Powers's "fact"-heavy approach led him to conclude that questions about the potential of artificial rainmaking could "only be settled by means of experiments" (93). As for Dyrenforth's perspiration thesis, it seems he derived this conclusion from his reading of Plutarch. See Dyrenforth, "Can We Make It Rain?," 387.

9. Dyrenforth, "Can We Make It Rain?," 393–94.

10. According to Cynthia Barnett, only two reporters were actually on the scene, one from the *Chicago Farm Implement News* and another from the *Dallas Farm and Ranch*. Additional newspapers obtained information by telegraph and other means. News about the Midland experiment traveled through circuitous channels, setting the stage for a contentious debate about the project's results and implications. See Cynthia Barnett, *Rain: A Cultural and Natural History* (New York: Crown, 2015), 164.

11. Dyrenforth, "Can We Make It Rain?," 395–96.

12. The first newspaper quotation is from the *Saint Louis Republic*, cited in the *Phillips County (KS) Inter Ocean*, August 1891, KSHS–"Rain and Rainfall" Clippings. 551.57R. The Moses quotation is from the *Chicago Times*, cited in Whitaker, "Making War," 211.

13. Though Dyrenforth perceived his experiment as a success, some observers immediately expressed their doubts. See Fleming, *Fixing the Sky*, 67. Historians including Fleming, Courtwright, Barnett, Harper, and Whitaker have provided more thorough accounts of the events of August 1891. My aim in reexamining the experiment and its contentious aftermath is to elucidate the link between rainmaking experiments and unsettled theoretical debates about human-induced climate change.

14. "The Rain-Makers," *Monthly Weather Review of the Iowa Weather and Crop Service* 2, no. 8 (August 1891).

15. A. W. Greely, "Some Peculiarities of the Rainfall of Texas," Philosophical Society of Washington, Bulletin 12 (Washington, DC: Philosophical Society, April 1892), 65.

16. "The Rain-Maker Here," unattributed 1892 newspaper clipping in "Records Relating to the Rain-Producing Experiment, 1892," National Archives, College Park, entry 123, box 1.

17. Courtwright, "On the Edge of the Possible," 544.

18. "Man-Made Rain," *New York Times*, August 28, 1892; clipping filed in "Records Relating to the Rain-Producing Experiment, 1892," National Archives, College Park, RG 95, entry 123, box 1. In the months leading up to the Midland experiment, Fernow did his utmost to avoid association with the Dyrenforth experiment. See Harold K. Steen, *The US Forest Service, a History* (Seattle: University of Washington Press, 1976), 44.

19. William Morris Davis, "The Theories of Artificial and Natural Rainfall," *American Meteorological Journal* 8, no. 1 (March 1892), 493 for "congressional action" and "sincere belief," 499 for "parody." An 1892 letter from Henry Holdes of Yuma, Colorado, to Secretary of Agriculture J. M. Rusk sheds light on how settlers and agriculturalists perceived rainmaking experiments. Holdes described himself as a "poor farmer" with "only a common school education." Despite his self-effacement, Holdes was very perceptive and had read reports by H. C. Dunwoody about percussion-induced rainfall. The Coloradan doubted theories ascribing storms to artillery barrages, but he believed explosive-

carrying balloons could be effective when used in favorable atmospheric conditions. If society could find a way to "equalize the weather," Holdes wrote, the rich soils near his home in northeastern Colorado could be transformed into nearly ideal agricultural territory. Holdes's letter is held in the "Records Relating to the Rain-Producing Experiment, 1892," National Archives, College Park, RG 95, entry 123, box 1.

20. For an early report claiming that Curtis's observations confirmed Dyrenforth's success, see the August 1891 *Phillips County (KS) Inter Ocean*, Kansas State Historical Society, "Rain and Rainfall" Clippings, 551.57R. The following year, some continued to invoke Curtis in their defenses of Dyrenforth. Edward Powers's August 1892 piece in the *American Meteorological Journal* cites a *New York Independent* article by Curtis asserting that the experiments created rain "a sufficient number of times to make it quite probable that the relation was that of cause and effect." As the author of *War and the Weather*—the chief inspiration for the Dyrenforth experiment—Powers stood to gain from Curtis's endorsement of the Midland tests. See Edward Powers, "Artificial Rain," *American Meteorological Journal* 9, no. 4 (August 1892): 180.

21. George E. Curtis, "The Facts about Rain-Making," *Engineering Magazine*, "Records Relating to the Rain-Producing Experiment, 1892," National Archives, College Park, RG 95, entry 123, box 1, 540.

22. George E. Curtis, "The Facts about Rain-Making," *American Meteorological Journal* 9, no. 6 (October 1892): 277–78.

23. Curtis, "Facts about Rain-Making," *Engineering Magazine*, 547–48. The meteorologist made similar claims in his October 1892 article in the *American Meteorological Journal*, adding that the artificial showers lasted only twenty to thirty seconds.

24. Curtis describes the brief artificial showers as an "interesting result" in "Facts about Rain-Making," 276.

25. Mark W. Harrington, "Weather-Making, Ancient and Modern," *Annual Report of the Board of the Regents of the Smithsonian Institution, 1894* (Washington, DC: GPO, 1896), 258.

26. Harrington, "Weather-Making," 265. In a pre-experiment letter to Fernow, Harrington wrote that he was "extremely doubtful" that Dyrenforth could produce rain. His doubts, however, did not prevent him from arguing to Fernow that the experiment "can be made of serious scientific value." See the September 30, 1890, letter from Mark W. Harrington to Bernhard Fernow, "Records Relating to the Rain-Producing Experiment, 1892," National Archives, College Park, RG 95, entry 123, box 1.

27. Harrington, "Weather-Making," 259. For a discussion of the problems and implications of twentieth- and twenty-first-century rainmaking and geoengineering, see Fleming, *Fixing the Sky*, 225–68, and Holly Jean Buck, "Geoengineering: Re-making Climate for Profit or Humanitarian Intervention?" *Development and Change* 43, no. 1 (2012): 253–70.

28. Barnett, *Rain*, 171; see also 159.

29. Lovewell, a respected figure in Great Plains meteorology, cited the battle-induced rainfall theory espoused by Powers and Dyrenforth in an 1892 piece. Although he called on rainmakers to offer more tangible proof of their theories, Lovewell allowed for the possibility that battles had caused "temporary climatic phenomena." See J. T. Lovewell, "Human Agency in Changing or Modifying Climate," *Quarterly Report of the Kansas State Board of Agriculture for the Quarter Ending March 31, 1892* (Topeka, KS: Hamilton Printing Company, 1892), 140; see also 143.

30. Barnett, *Rain*, 165. One of the most prominent post-Dyrenforth rainmakers was

Charles William Post, of breakfast cereal fame. In the early 1910s, Post carried out rain-making experiments on his vast planned community in West Texas. For more on Post, see Courtwright, "On the Edge of the Possible," 546, and Whitaker, "Making War," 214–15.

Chapter 5

1. *Topeka Commonwealth*, May 21, 1885. Philip J. Pauly shows how nineteenth-century Americans viewed various "cultures"–horticulture, arboriculture, silviculture, agriculture—as connected to broader "culture." See Philip J. Pauly, *Fruits and Plains: The Horticultural Transformation of America* (Cambridge, MA: Harvard University Press, 2007), 6.

2. Anne André-Johnson, *Notable Women of St. Louis, 1914* (St. Louis, MO: Woodward, 1914), 59.

3. S. W. Dodds, M.D., "What Causes the Cyclones," *Proceedings of the American Forestry Association at the Summer Meeting, Held in Quebec, September 2–5, 1890, and at the Ninth Annual Meeting, Held at Washington, December 30, 1890* (Washington DC, 1891), 99–101.

4. *Southern Lumberman*, August 1, 1896, Forest History Society, USFS Clippings File, box 45.

5. *Topeka Commonwealth*; Dodds, "What Causes the Cyclones," 101.

6. Laura J. Martin, *Wild by Design: The Rise of Ecological Restoration* (Cambridge, MA: Harvard University Press, 2022).

7. Donald Worster, *Nature's Economy: A History of Ecological Ideas* (1977; repr., Cambridge: Cambridge University Press, 1985), xiv.

8. Conevery Bolton Valenčius, *The Health of the Country: How American Settlers Understood Themselves and Their Land* (New York: Basic Books, 2002), 67–70, 99; David Lowenthal, *George Perkins Marsh: Prophet of Conservation* (Seattle: University of Washington Press, 2000), 292. For the origins of American environmental sensibilities, see also Aaron Sachs, *The Humboldt Current: Nineteenth-Century Exploration and the Roots of American Environmentalism* (New York: Viking, 2006), 32. For equilibrium ecology within forestry, see Emily Brock, "The Challenge of Reforestation: Ecological Experiments in the Douglas Fir Forest," *Environmental History* 9, no. 1 (2004): 62.

9. For an exploration of the many meanings of trees in nineteenth-century culture, see Daegan Miller, "Reading Tree in Nature's Nation: Toward a Field Guide to Sylvan Literacy in the Nineteenth-Century United States," *American Historical Review* 121, no. 4 (October 2016): 1114–40.

10. Scholars of American forestry have sometimes dismissed the ecological paradigm of these theorists either as "wrong" or as a failed amateurish precursor to scientific forestry. See Michael Williams, *Americans and Their Forests: A History of Geography* (Cambridge: Cambridge University Press, 1989), 381; for "wrong," see Donald J. Pisani, "Forests and Conservation, 1865–1890," in *American Forests: Nature, Culture, and Politics*, ed. Char Miller (Lawrence: University of Kansas Press, 1997), 26. My approach builds on the work of Nancy Langston and Robert Gardner, historians who seriously considered the ecological approach of nineteenth-century foresters. See Langston, *Forest Dreams, Forest Nightmares: The Paradox of Old Growth in the Inland West* (Seattle: University of Washington Press, 1995), 144–45; Gardner, "Constructing a Technological Forest: Nature, Culture, and Tree-Planting in the Nebraska Sand Hills," *Environmental History* 14, no. 2 (April 2009): 276.

11. "In Memory of John A. Warder," Forest History Society, George H. Writ Collection, Warder Biography and Obituary File, folder 17 (unattributed).

12. For a characterization of Warder as one of the last proponents of an older tradition, see Andrew Denny Rodgers III, *Bernhard Eduard Fernow: A Story of North American Forestry* (Durham, NC: Forest History Society, 1991), 27.

13. Biographical details on Warder are from Rodgers, *Fernow*, 48–49.

14. Nineteenth-century forestry writings are rife with debates about the possibility of a timber famine, though not all agreed about the scope or imminence of the threat. In 1885, for example, the American Forestry Congress issued a stark warning: "Without joining in the cries of the alarmists, we have good reasons and sufficient data to assert that the present policy, if continued, must seriously affect this factor of national wealth at no distant time." See *Proceedings of the American Forestry Congress at Its Meeting Held in Boston, September, 1885* (Washington, DC: Judd and Detweiler, 1886), 4. Massachusetts politician George Loring, on the other hand, dismissed doomsday warnings of timber scarcity as alarmist. See Loring's "Speech on Forestry" before the American Forestry Congress, Boston, 1885, George B. Loring Papers, Peabody-Essex Museum, box 8, folder 12.

15. John A. Warder [member of the Scientific Commission of the United States—Vienna International Exposition, 1873], *Report on Forests and Forestry* (Washington, DC: GPO, 1875), 10. European foresters may have presented Warder with an overly rosy interpretation of the successes of their forestry. James Scott has studied the unintended consequences and ecological crises precipitated by Prussian forestry. See Scott's chapter "Nature and Space" in his *Seeing Like a State: How Certain Schemes to Improve the Human Condition Have Failed* (New Haven, CT: Yale University Press, 1998), 11–52. For more on transatlantic influences in forest culture and restoration, see Marcus Hall, *Earth Repair: A Transatlantic History of Environmental Restoration* (Charlottesville: University of Virginia Press, 2005).

16. Note on terminology: "afforestation" generally refers to planting initiatives in treeless, arid, or long-since-deforested areas; "reforestation" is the effort to restore forests to recently clear-cut areas; "silviculture" or "forest culture" denotes the broader cause of forest planting, including efforts to instill respect for trees and forests.

17. *Proceedings of the American Forestry Congress at Its Meeting Held in Boston, September, 1885* (Washington, DC: Judd and Detweiler, 1886), 4.

18. John A. Warder, "Larch Wood," in *The American Journal of Forestry*, ed. Franklin B. Hough, vol. 1, September 1882–October 1883 (Cincinnati, OH: Robert Clarke, 1882–83), 14.

19. "Tree planting is indeed a necessity" and "great question" are both from Warder's "Address Delivered before the Otoe County Horticultural Society in Nebraska City, September 12th, 1878," "The Future Orchards and Forests of Nebraska," Nebraska State Historical Society, 634.9 W21a, 3, 5.

20. John Warder, "Tree Planting in Shelter Belts," *Journal of the American Agricultural Association*, May 1882, 1–4.

21. Langston, *Forest Dreams, Forest Nightmares*, 99.

22. Warder, "Future Orchards," 7.

23. Samuel Hays's exhaustive study of Progressive Era conservationism characterized the movement's guiding spirit as "the gospel of efficiency." Samuel P. Hays, *Conservation and the Gospel of Efficiency: The Progressive Conservation Movement, 1890–1920* (1959; repr., New York: Atheneum, 1972).

24. For an example of Warder's advocating for "practical knowledge," see Warder, "Tree Planting in Shelter Belts," 1. For an example of Warder's pushing for centralization and expertise, see John Warder, "Some Trees for Planting on the Open Prairies of Northern Illinois and Adjoining Regions," *Transactions of the Illinois State Horticultural Society*, 1881, 2–3.

25. John Warder, "Tree Planting for Railroads," read at the Montreal Meeting of the American Forestry Congress, 1882, 1, Forest History Society, Hirt Collection, folder 15. Jared Farmer has shown how profit-driven tree planting schemes captured the imagination of investors in California from the 1870s to the early 1900s. Warder espoused a starkly different form of tree advocacy than did California planters and other figures who sought "instant industrial forests." See Jared Farmer, *Trees in Paradise: A California History* (New York: W. W. Norton, 2013), 138.

26. John Warder, "What Are Forest Trees?," printed at the request of the secretary of the Kansas Horticultural Society and submitted to the commissioner of the General Land Office for "the purpose of securing a reconsideration of the ruling of the Department which excluded the Catalpa, Osage Orange, and Ailantus from being planted under the timber acts of Congress" (North Bend, OH, January 10, 1882).

27. Warder, "Tree Planting for Railroads," 2–3.

28. Benjamin Gott, "Forest Tree-Planting: The Results and Advantages to Farmers," in *The American Journal of Forestry*, ed. Franklin B. Hough, vol. 1, September 1882–October 1883 (Cincinnati, OH: Robert Clarke, 1882–83), 339–47. Canadian foresters and forestry advocates played a large role in debates about forestry, conservation, and climate. Deforestation around the Great Lakes and Saint Lawrence River and its supposed climatic consequences galvanized Gott and like-minded authors. See John L. Riley, *The Once and Future Great Lakes Country: An Ecological History* (Montreal: McGill University Press, 2013), 186–89. Further west, Canadian railroads implemented climate-motivated experimental plantations similar to those established in Kansas and Nebraska. See H. G. Joly de Lotbinière, "Tree Planting on the Prairies," *Proceedings of the American Forestry Association at the Summer Meeting, Held in Quebec, September 2–5, 1890 and at the Ninth Annual Meeting, Held at Washington, December 30, 1890* (Washington, DC, 1891), 68–69.

29. Leonard B. Hodges, "The Planting of Wind-Breaks along Rail-Roads," in *Journal of Forestry*, 252.

30. H. M. Thompson, "Plan of Forest Planting for the Great Plains of North America," in *Journal of Forestry*, 229–30. In 1888 the University of Pennsylvania professor E. J. James made a similar argument, highlighting the sometimes contentious semantics of forest culture. James wrote: "I would emphasize the fact that tree-planting is not forest culture." For James, tree planting was carried out piecemeal while forest culture was more systematic. He favored the latter, arguing that government should play a stronger role in preventing haphazard, profit-minded laissez-faire planting. See E. J. James, "The Government in Its Relation to the Forests," Department of Agriculture, Forestry Division, Bulletin 2 (Washington, DC: GPO, 1888), 29–30.

31. Bruno Latour, *The Politics of Nature: How to Bring the Sciences into Democracy*, trans. Catherine Porter (Cambridge, MA: Harvard University Press, 2004), 21, 25.

32. In discussing the relation between uncertainty and the utopia-dystopia dialectic, I am drawing from Michael Golding, Helen Tilley, and Gyan Prakash, *Utopia/Dystopia: Conditions of Historical Possibility* (Princeton, NJ: Princeton University Press, 2010). In the introduction, the authors write that "utopia, dystopia, chaos: these are not just ways

of imagining the future (or the past) but can also be understood as concrete practice through which historically situated actors seek to reimagine their present and transform it into a plausible future" (2).

33. Cassius M. Clay, "The Preservation of Forests," in *The American Journal of Forestry*, 1:462–63. For a biographical study of Clay, see David L. Smiley, *Lion of Whitehall: The Life of Cassius M. Clay* (Madison: University of Wisconsin Press, 1961).

34. Helen Tilley, "Ecologies of Complexity: Tropical Environments, African Trypanosomiasis, and the Science of Disease Control in British Colonial India," *Osiris* 19 (2004): 33; see also 26.

35. Eric Rutkow gives a brief overview of Morton's political leanings in *American Canopy: Trees, Forests, and the Making of a Nation* (New York: Scribner, 2012), 132.

36. Furnas "respectfully dedicated" his book *Arbor Day* (Lincoln, NE: State Journal Company, 1888) to Morton. For evidence of collaboration between Morton and Furnas in promoting Nebraska, see February 3 and February 5, 1873, letters between Morton and Furnas in the Furnas Biographical File, Nebraska State Historical Society.

37. Warder, "Future Orchards," 3.

38. For information on Morton's background, see Morton Biographical File, Nebraska State Historical Society, RG 1012; see also Nebraska State Historical Society, J. Sterling Morton Pamphlet Collection, vol. 9, 080 M84; and Rutkow, *American Canopy*, 130–33.

39. See James C. Olsen, "Arbor Day—A Pioneer Expression of Concern for Environment," *Nebraska History* 53 (1972). Michael Williams characterizes Morton's environmental thought as "eclectic and confused" while noting the Arbor Day movement's "ethos of mysticism" and "quasi-religious" rhetoric. See Williams, *Americans and Their Forests*, 383. Philip Pauly's *Fruits and Plains: The Horticultural Transformation of America* (Cambridge, MA: Harvard University Press, 2007) offers a brief overview of Arbor Day in a chapter on silviculture and horticulture on the Great Plains.

40. In his tome *The Mississippi Valley* (Chicago: S. C. Griggs, 1869), J. W. Foster described the strange effect of the Plains on people "accustomed to look out upon a landscape diversified by mountain and valley" where "every hill had its crown of forest, and every stream its waterfall" (72). Not all settlers shared this inclination, however. In 1856 Kansas settler Sara Robinson preferred the stark prairie landscape, which "nature had made singularly beautiful." See Sara T. L. Robinson, *Kansas: Its Interior and Exterior Life, Including a Full View of Its Settlement, Political History, Social Life, Climate, Soil, Productions, Scenery Etc.* (Boston: Crosby, Nichols, 1856), 2–3. Over the course of his western travels, J. H. Beadle claimed he "never saw a farmer's wife who had tried both, who did not prefer the prairie to the timber, despite the intense cold of winter." See J. H. Beadle, *The Undeveloped West, or Five Years in the Territories* (St. Louis, MO: National Publishing Company, 1873), 40. For a discussion of gender and perceptions of the prairie landscape, see Courtney Wiersema, "A Fruitful Plain: Fertility on the Shortgrass Prairie," *Environmental History* 16, no. 4 (2011): 678–99.

41. See Morton, "Arbor Day" (1886), 49.

42. Furnas, *Arbor Day* (1888), 7.

43. B. G. Northrop, "Forests and Floods," in *Report of Secretary of Connecticut Board of Agriculture* (Hartford, CT: Case, Lockwood, and Brainerd, 1885), 18.

44. J. Sterling Morton, "Arbor Day," in *Proceedings of the American Forestry Congress at Its Meeting Held in Boston, September, 1885* (Washington, DC: Judd and Detweiler, 1886), 50.

45. Many members of the East-centric Forestry Congress and Forestry Association adopted Arbor Day as their own. See *Proceedings of the American Forestry Congress* (1885), 4–5. For the international spread of Arbor Day, see Rutkow, *American Canopy*, 133.

46. Morton, "Arbor Day" (1886), 50–51.

47. Linda Nash, *Inescapable Ecologies: A History of Environment, Disease, and Knowledge* (Berkeley: University of California Press, 2006), 50 for "assumed separation" and 73–74 for "asserted the agency." Nash also writes that "for nineteenth-century Americans, the body itself was not a clearly bounded entity, separate and distinct from its surroundings" (24). See also Gardner, "Constructing a Technological Forest," 288.

48. Morton, "Arbor Day: Its Origin and Growth," Address of J. Sterling Morton Delivered April 22, 1886, at the State University, Lincoln, NE, Nebraska State Historical Society, J. Sterling Morton Pamphlet Collection, vol. 70.

49. In his history of North Atlantic fisheries, Jeffrey Bolster illustrates how, during the Gilded Age, laissez-faire economic interests deployed the notion of ecological unknowns. Despite signs of overfishing, Bolster explains, "oil and guano interests insisted that nothing untoward was happening; after all, the ocean produced fish in 'natural' ways largely unknowable to humans, and in ways—they believed—that should be beyond the compass of law." See Bolster, *The Mortal Sea: Fishing the Atlantic in the Age of Sail* (Cambridge, MA: Harvard University Press, 2012), 181.

50. See Rutkow, *American Canopy*, 131.

51. N. H. Egleston, *Arbor Day Leaves: A Complete Programme for Arbor Day Observance, Including Readings, Recitation, Music, and General Information* (New York: American Book Company, 1893). Nebraska State Historical Society, J. Sterling Morton Pamphlet Collection, vol. 70.

52. John Peaslee, "Trees and Tree Planting, with Exercises and Directions for the Celebration of Arbor Day," in *Planting Trees in School Grounds and the Celebration of Arbor Day*," Department of the Interior, Bureau of Education (Washington, DC: GPO, 1885), 52, Nebraska State Historical Society, J. Sterling Morton Pamphlet Collection, vol. 63. Memorials such as those described by Cary fit within the Arcadian landscape tradition chronicled by Aaron Sachs in *Arcadian America: The Death and Life of an Environmental Tradition* (New Haven, CT: Yale University Press, 2013).

53. John B. Peaslee, "Arbor Day or Tree-Planting Celebration?," *Proceedings of the American Forestry Congress at Its Meeting Held in Boston, September, 1885* (Washington, DC: Judd and Detweiler, 1886), 48–49. The quotation about the "future of this great republic" is printed in Peaslee's pamphlet, but it is unclear if it is by Peaslee himself or if Peaslee is quoting Emil Rothe of the American Forestry Congress.

54. Furnas, *Arbor Day* (1888), 23.

55. For a discussion of class tensions in the context of agricultural education, see Robert P. Crawford, *These Fifty Years: A History of the College of Agriculture* (Lincoln: University of Nebraska College of Agriculture, 1925); see especially 18 for suspicion of "book farming." See also Pisani, "Forests and Conservation," 27.

56. Diana K. Davis has examined the role of afforestation initiatives in legitimizing and bolstering colonial settlement projects. Davis, *Resurrecting the Granary of Rome: Environmental History and French Colonial Expansion in French North Africa* (Athens: Ohio University Press, 2007). 78. See also Davis, *The Arid Lands: History, Power, Knowledge* (Cambridge, MA: MIT Press, 2016).

57. For "growing tendency," see Hough and Emerson, New York State Library, box 39,

folder 1 (1874), p. 2. The quotation about "repose" is from the New York State Library's guide to the Franklin B. Hough Papers; see http://www.nysl.nysed.gov/msscfa/sc7009 .htm. For Hough's discussion of how human agency can influence extreme weather events (or "vicissitudes of climate," as he terms them), see Franklin B. Hough. *Report upon Forestry* (Washington, DC: GPO, 1878), 221.

58. Thomas Meehan, "Forests the Result and Not the Cause of Climate," *Prairie Farmer* 44, no. 47 (November 22, 1873).

59. Hough was born in Martinsburg, New York, in 1822. First as a doctor and later as an author and state and federal employee, Hough traveled across New York and lived in several towns throughout his home state, including Somerville, Albany, and Lowville. He recorded daily weather observations during his travels—including temperature readings, precipitation totals, cloud cover, and wind direction—in his meteorological journals. For an example of Hough's meteorological journals, see Franklin B. Hough Papers, New York State Library, box 8. For examples of Hough's later data collection efforts from his time working as a state agent, see NYSL, box 42. "Concert of observation" quotation is from Hough's 1856 "Annual Report" to the Assembly of the State of New York State; see NYSL, box 9, folder 3.

60. See Williams, *Americans and Their Forests*, 400; Rodgers, *Fernow*, 27; Ramachandra Guha, *Environmentalism: A Global History* (New York: Longman, 2000), 30; for state governance and data collection in the Unites States, see Susan Schulten, *Mapping the Nation: History and Cartography in Nineteenth-Century America* (Chicago: University of Chicago Press, 2012).

61. See railroad circulars in New York State Library, Hough Papers, box 40. Hough wanted to know if and how railroads attempted to reduce extensive use of timber resources during their westward expansion.

62. For Hough on waste-land, see "The Value of American Timber Lands," in *The Proper Value and Management of Government Timber Lands and the Distribution of North American Forest Trees*, Department of Agriculture, Miscellaneous, Special Report no. 5 (Washington, DC: GPO, 1884), 9.

63. Hough Papers, New York State Library, box 49.

64. For Hough on "Arbor Days," see "Value of American Timber Lands" (1884), 6.

65. Hough Papers, New York State Library, box 9, folder 1, 4.

66. Franklin B. Hough, "Essay on the Climate of the State of New York," prepared at the request of the Executive Committee of the State Agriculture Society and published in the fifteenth volume of their *Transactions* (Albany, NY: Van Benthuysen, 1857), 31. For examples of halos and Hough's daily weather observations, see New York State Library, box 8, folder 1.

67. Williams, *Americans and Their Forests*, 400.

68. Hough, manuscript for *Meteorology and Climate of New York* (1863), New York State Library, box 9, folder 1, n.p. For a discussion of Boussingault's observations about deforestation-induced desiccation in South America, see Gregory Cushman, *Guano and the Opening of the Pacific World: A Global Ecological History* (Cambridge: Cambridge University Press, 2013), 36.

69. Determining who first advanced these proto-ecological notions is in many ways a fool's errand. Marsh published several proto-ecological tracts before *Man and Nature* (see, for example, "The Study of Nature," *Christian Examiner*, January 1860), but so did Hough (see his 1857 *Essay on the Climate of the State of New York*). In my view, the important thing is not who first devised a version of forest-climate ecology, but that Hough

and Marsh both belonged to a sociocultural milieu comprising many thinkers interested in questions of natural interdependence.

70. See Hough Papers, New York State Library, box 9, folder 1.

71. Lowenthal, *Prophet of Conservation*, 303 for Hough's views on Marsh and 508n29 for Marsh's dismissal of Hough.

72. See Worster, *Nature's Economy*, 172–73, 185. Though Marsh viewed humanity as a great "disturbing agent," he also believed that humans did not fully understand the natural systems they were transforming. See Lowenthal, *Prophet of Conservation*, 278, 426.

73. Hough Papers, New York State Library, box 9, folder 1.

74. Deborah Coen, "Imperial Climatographies from Tyrol to Turkestan," *Osiris* 26 (2011): 45–65.

75. See David Emmons, *Garden in the Grasslands: Boomer Literature of the Central Great Plains* (Lincoln: University of Nebraska Press, 1971), x, 6–10. Emmons uses a quotation from Thomas Carlyle to encapsulate the belief in human dominance over nature: "Nothing can resist us [the Anglo-Saxons]. We war with rude nature; and, by our resistless engines, come off always victorious, and loaded with spoils."

76. F. H. Snow, "Is the Rainfall of Kansas Increasing?," *Kansas City Review of Science and Industry*, vol. 8 (Kansas City, MO: Ramsey, Millet, and Hudson, 1885), 458.

77. Ellwood Cooper, *Forest Culture and Eucalyptus Trees* (San Francisco, CA: Cubery, 1876), 10.

78. *Garden City (KS) Herald*, March 24, 1883, Kansas State Historical Society Newspapers. For the record, Garden City had approximately 18,000 inhabitants in 1980, so the newspaper was more than a little optimistic.

79. Adolphus Greely, manuscript on "Mississippi Floods," Adolphus Greely Papers, Library of Congress, box 82 (MS is dated "90"), 1.

80. Samuel Aughey, "The Increasing Need for Forests," *Nebraska Farmer* 1, no. 10 (October 1877).

81. Pauly describes this debate as part of a broader effort to "fix the grasslands," a landscape unfamiliar to Euro-Americans. See Pauly, *Fruits and Plains*, 89–90. According to Pisani, the staunchest proponents of forestry viewed all treeless areas as aberrant and in need of correction. He describes the forester Bernhard Fernow as believing that "the entire earth would have been covered with forests, 'save only a few localities,' had man and animals not interfered." Similarly, in 1891 Fernow stated that "the entire earth is a potential forest." See Pisani, "Forests and Conservation," 19. For an analysis of the ancient Great Plains in the Euro-American cultural and geological imagination, see Daniel Zizzamia, "Restoring the Paleo-West: Fossils, Coal, and Climate in Late Nineteenth-Century America," *Environmental History* 24, no. 1 (2019): 130–56.

82. "Letter from Dr. Franklin B. Hough, in Regard to the Effect of Forests in Increasing the Amount of Rainfall," H.R., 48th Cong., 2nd Sess., Ex. Doc. (Washington, DC: GPO, 1885), app. 14, 131.

83. Marcus Hall explains how restoration discourses rely on imagined pasts in *Earth Repair*, 127–28.

84. For settlers' attempts to come to terms with prairie and plain ecologies, see Wiersema, "Fruitful Plain."

85. Henry Inman, "On Climatic Changes in the Prairie Region of the United States," in *Kansas City Review of Science and Industry*, vol. 2 (Kansas City, MO: Ramsey, Millet, and Hudson, 1885), 236–37.

86. J. K. Macomber, "Adaptability of Prairie Soils for Timber Growth," *Transactions of the Iowa State Horticultural Society for 1879* (Des Moines, IA: F. M. Mills, 1880) 293.

87. Edgar T. Ensign, "Report on the Forest Conditions of the Rocky Mountains," US Department of Agriculture, Forestry Division, Bulletin 2, *Report on the Forest Conditions of the Rocky Mountains, and Other Papers* (Washington, DC: GPO, 1888), 82–83.

88. J. D. Whitney, "Plain, Prairie, Forest," *American Naturalist* 10, no. 10 (October 1876): 577.

89. Hough, "Report upon Forestry" (1878), 221.

90. Hough, "Elements of Forestry" (1882), 21–22.

91. Hough, "Elements of Forestry" (1882), 17–18.

92. *Letter from Dr. Franklin B. Hough.*

93. John H. Klippart, "Forests: Their Influence upon Soil and Climate," *Fifteenth Annual Report of the Ohio State Board of Agriculture, for the Year 1860* (Columbus, OH: Richard Nevins, 1861), 255–57.

94. John Hay, "Atmospheric Absorption and Its Effect upon Agriculture," *Proceedings of the Eighteenth Annual Meeting 1890* of the Kansas State Board of Agriculture (Topeka, KS, 1890), 124 for "stomata," 126 for "stupendous."

95. For miasmas, see Klippart, "Forests," 259; for reservoirs, see Hay, "Atmospheric Absorption," 126. For a discussion of high modern climate engineering schemes in the turn-of-the-century context, see Lehmann, "Infinite Power to Change the World: Hydroelectricity and Engineered Climate Change in the Atlantropa Project," *American Historical Review* 121, no. 1 (February 2016): 70–100.

96. For an example of Hough's hesitation when faced with scientific uncertainty and contentious debates, see Hough, "Report upon Forestry" (1878), 221. Hough urged caution when implementing policy, writing that "conditions of climate should be understood before forest-cultivation is attempted."

97. Bernhard Fernow, "The Influence of Forests on Irrigation Problems," Fernow Papers, Cornell University Library, box 2, folder 25, 1. The folder heading dates the manuscript as being written between 1885 and 1900.

98. Fernow, "Forest as a Condition of Culture," Fernow Papers, box 2, folder 23, undated manuscript; the exact date is unclear, but the folder heading attributes the paper to either 1885–88 or 1892. In her study on forestry in the Blue Mountains of Oregon and Washington, Nancy Langston credits Fernow for acknowledging complex ecological interconnections and their mysteriousness: "He saw the forest as a complex whole whose functions people could only dimly understand." See Langston, *Forest Dreams, Forest Nightmares,* 107.

99. Fernow, "What should be the attitude which society, the community, the state should take toward forestry," Fernow Papers, Cornell University, box 2, folder 27; dated only as 1885–1900.

100. Rodgers, *Fernow,* 14–17.

101. Fernow's institutional struggles are exhaustively chronicled by Rodgers in the later chapters of his biography.

102. Fernow, "What Should Be the Attitude . . . ," Fernow Papers, Cornell University, box 2, folder 27 (1885–1900).

103. Fernow, "Forests as a Condition of Culture," Fernow Papers, box 2, folder 23.

104. Fernow Papers, box 2, folder 27.

105. For Fernow on Hough, see Williams, *Americans and Their Forests,* 400. For Fernow on "friends" and "popular writers," see *Introduction and Summary of Conclusions,* 9.

106. R. Max Peterson writes that Fernow helped launch the "real forestry movement" in North America. Rodgers argues that Fernow led US forestry out of its folk-

loric, unprofessional, propagandistic beginnings. See Peterson's foreword to Rodgers's biography (xi-xii) and Rodgers, *Fernow*, 13, 27. Donald Pisani, by contrast, points out the overlaps and continuity between Hough's and Warder's brand of "forest culture" and Progressive Era forest management. See Pisani, "Forests and Conservation," 26.

107. Rogers, *Fernow*, 71 for "true science" and 364 for the argument about ecology.

108. Environmental historians and other scholars have blurred the dichotomy between the Gilded Age and the Progressive Era, arguing that concerns about the excesses of unchecked capitalism predated the advent of high Progressivism. See, for example, Leon Fink, *The Long Gilded Age: American Capitalism and the Lessons of a New World Order* (Philadelphia: University of Pennsylvania Press, 2015), 2–4; Ellen Stroud, *Nature Next Door: Cities and Trees in the American Northeast* (Seattle: University of Washington Press, 2012), 30.

109. Pisani, "Forests and Conservation," 26.

110. See Klippart, "Forests," 260–61. Rodgers writes that Fernow's overarching goal was "the exact ascertainment of the scope and nature of . . . indirect beneficial influences on the life of man." See Rodgers, *Fernow*, 30.

111. Fernow Papers, box 2, folder 28.

112. For a discussion of the gradual adoption of germ theory and the persistence of some forms of medical geography and miasma theory, see Nash, *Inescapable Ecologies*, 83.

113. For folklore and vernacular science, see Valenčius, *Health of the Country*, 22, 76. For an example of earlier "forest culture" beliefs that prefigured Fernow's writings, see C. A. Logan, "Report on the Sanitary Conditions of the State of Kansas" (1865), University of Kansas, Spencer Library, RH C5773, 154–55. Logan's discussion of urban forestry made arguments strikingly similar to Fernow's.

114. Fernow Papers, box 2, folder 28. For an analysis of the malaise caused by turn-of-the-century industrialization and urbanization, see T. J. Jackson Lears, *No Place of Grace: Antimodernism and the Transformation of American Culture, 1880–1920* (New York: Pantheon, 1981).

115. Loring, "Speech on Forestry" (1885), 15–16.

116. Tiffin Sinks, "Report on the Climatology of Kansas" (1865), University of Kansas, Spencer Library, RH C5773, 185.

117. Fernow Papers, box 2, folder 25. Fernow also wrote that "climatic conditions are in the first place due to cosmic, unterrestrial influences" (12).

Chapter 6

1. For a discussion of "scientific dry farming," see Gary D. Libecap and Zeynep K. Hansen, "Rain Follows the Plow and Dryfarming Doctrine: The Climate Information Problem and Homestead Failure in the Upper Great Plains," *Journal of Economic History* 62, no. 1 (March 2002): 96–102.

2. "Is the Dry Climate of Central Kansas Changing?," *Topeka (KS) Daily Capital*, May 7, 1905.

3. For biographical details on Moore, see Jamie L. Pietruska, "US Weather Bureau Chief Willis Moore and the Reimagination of Uncertainty in Long-Range Forecasting," *Environment and History* 17, no. 1 (February 2011): 79–105. Pietruska writes that Moore believed in the importance of "bureaucratic orchestration" and the "professional

expertise of government meteorologists (80). For a detailed analysis of technocracy, "techno-politics," and the ability of experts to conjure and deploy scientific knowledge, see Timothy Mitchell, *Rule of Experts: Egypt, Techno-Politics, Modernity* (Berkeley: University of California Press, 2002). Following Mitchell, I view technocratic governance as a political project that derives its legitimacy from technoscientific expertise.

4. For "speculators and land boomers," see the *Topeka Daily Capital*, February 19, 1907. For "we should not be misled," see the *Kansas City Journal*, February 1907, in "Rain and Rainfall" Clippings, Kansas State Historical Society, 551.57R.

5. In addition to newspapermen and agriculturalists, one of Kansas's preeminent climate experts added his voice to the anti-Moore chorus. F. H. Snow of the University of Kansas wrote a piece rebuking Moore's claims. See F. H. Snow, "Change in the Climate of Kansas," *Transactions of the Kansas Academy of Science* 20, part 2 (Topeka, KS: State Printing Office, 1907).

6. *Topeka Daily Herald*, January 1907, in "Rain and Rainfall" Clippings, Kansas State Historical Society, 551.57R. "Rocky mountain canary" was a term used for donkeys.

7. *Kansas City Journal*, January 17, 1907. The editors of the *Journal* described Abernathy's letter as "forceful and logical," indicating that much of Kansas's press supported the reaction against Moore.

8. *Kansas City Journal*, January 18, 1907. Walter Kollmorgen's work shows that the clash between ranchers and tree-planting advocates (along with their agriculturalist allies) continued to rage into the "dry farming" era of the 1900s. See Kollmorgen, "The Woodsman's Assaults on the Domain of the Cattleman," *Annals of the Association of American Geographers* 59, no. 2 (June 1969): 233–38.

9. *Kansas City Journal*, January 28, 1907.

10. For a discussion of the role of windmills and irrigation in Great Plains agriculture, see Walter Prescott Webb, *The Great Plains* (1931; repr., Waltham, MA: Blaisdell, 1959), 336–66. For a more recent analysis of technologies of environmental improvement in the Great Plains, see David Nye, *America as Second Creation: Technology and Narratives of New Beginnings* (Cambridge, MA: MIT Press, 2003), 212–30. T. J. Jackson Lears describes the growing importance of race (and especially of scientific racism) in the Euro-American Progressive Era cultural imagination in *Rebirth of a Nation: The Making of Modern America* (New York: HarperCollins, 2009), 92–132.

11. *Kansas City Star*, January 26, 1907.

12. For an overview of the historiography on populism, see Joe Creech, "*The Tolerant Populists* and the Legacy of Walter Nugent," *Journal of the Gilded Age and Progressive Era* 14, no. 2 (April 2015): 141–59. See also Noam Maggor, *Brahmin Capitalism: Frontiers of Wealth and Populism in America's First Gilded Age* (Cambridge, MA: Harvard University Press, 2017).

13. *Topeka Daily Herald*, March 9, 1907.

14. *Kansas City Star*, June 9, 1912.

15. Karen De Bres writes that the end of the nineteenth century saw the "refutation" of "climate-change theories." See De Bres, "Come to the Champagne Air: Changing Promotional Images of the Kansas Climate, 1854–1900," *Great Plains Quarterly* 23, no. 2 (Spring 2003): 122. De Bres's statement is consistent with the interpretations of an earlier generation of scholars, including Charles Kutzleb, who argued that "by 1895 a combination of evidence that rainfall was not increasing, the rise of new models of farming, a series of disastrous droughts, and lessons learned from the experience of Plains residents ended the stubborn belief" in human-induced climate improvement.

See Charles D. Kutzleb, "Rain Follows the Plow: History of an Idea" (PhD diss., University of Colorado, 1968), 383. For other descriptions of the decline and disappearance of agriculture- or forest-based climate theories around the 1890s and the turn of the century, see Williams, *Americans and Their Forests*, 386; Pisani "Forests and Conservation," 25; Rutkow, *American Canopy*, 137. A few scholars have shown that climate theories remained significant into the twentieth century. Joel Orth's work illustrates that climate beliefs continued to influence afforestation initiatives into the later Progressive Era and during the New Deal. See Orth, "Directing Nature's Creative Forces: Climate Change, Afforestation, and the Nebraska National Forest," *Western Historical Quarterly* 42 (Summer 2011): 197–217. A recent article by Susan E. Swanberg shows that climate change theses continued to circulate into the 1900s despite growing skepticism about theories of "rain following the plow." See Susan E. Swanberg, "'The Way of the Rain': Towards a Conceptual Framework for the Retrospective Examination of Historical American and Australian 'Rain Follows the Plough/Plow' Messages," *International Review of Environmental History* 5 (2019): 67–95.

16. Such struggles have been amply chronicled by Louis Warren and Karl Jacoby. See Warren, *The Hunter's Game: Poachers and Conservationists in Twentieth-Century America* (New Haven, CT: Yale University Press, 1997); Jacoby, *Crimes against Nature: Squatters, Poachers, Thieves, and the Hidden History of American Conservation* (Berkeley: University of California Press, 2003). Warren, for example, describes "conflicts between centralization and decentralization, between expertise and localism" during a time when the federal government emerged as a "major, reorganizing force" in people's lives (175).

17. *Kansas City Journal*, January 28, 1907.

18. Bernard Mergen, *Weather Matters: An American Cultural History since 1900* (Lawrence: University of Kansas Press, 2008), 13.

19. See Pietruska, "US Weather Bureau Chief," 99. Pietruska views the year 1906 as the turning point in Moore's career and in his acceptance of uncertainty.

20. Willis L. Moore, "A Report on the Influence of Forests on Climate and on Floods," House of Representatives, United States Committee on Agriculture (Washington, DC: GPO, 1910), 3.

21. Moore, "Report on the Influence," 9.

22. Moore, "Report on the Influence," 5. Moore also claimed that petrified forests in the southwestern United States offered proof that desiccation took place irrespective of human influences and changes in forest cover: "Unmistakable evidence is found of the existence of extensive forests in Arizona and New Mexico, where only the petrified trunks of trees now remain" (7).

23. Ian Tyrrell, *Crisis of the Wasteful Nation: Empire and Conservation in Theodore Roosevelt's America* (Chicago: University of Chicago Press, 2015), 127.

24. For a discussion of the technocratic production of uncertainty during a later period, see Shiloh R. Krupar, *Hot Spotter's Report: Military Fables of Toxic Waste* (Minneapolis: University of Minnesota Press, 2013), 172–74, 195.

25. See Lears, *Rebirth of a Nation*, 239 and 310. Kevin Armitage makes a similar argument about the nature study movement, stressing the alienating quality of scientific management and characterizing the movement as an attempt to "embrace scientific modernity while simultaneously recoiling from the narrow, instrumental, and ugly society created by industrial civilization" (1). Like Lears, Armitage believes that Taylorism, "the scientific worldview," and the search for "maximum productivity and efficiency" had a profoundly alienating effect on the daily lives of turn-of-the-century Americans. See

Kevin C. Armitage, *The Nature Study Movement: The Forgotten Popularizer of America's Conservation Ethic* (Lawrence: University of Kansas Press, 2009), 2–3, 9, 109. For a classic study on Progressive Era thinking, see Samuel P. Hays, *Conservation and the Gospel of Efficiency: The Progressive Conservation Movement, 1890–1920* (Cambridge, MA: Harvard University Press, 1959).

26. Recent scholarship has worked to develop a more expansive notion of both progressivism and conservationism. In the wake of the interventions by Benjamin H. Johnson, Dorceta E. Taylor, Kevin Armitage, Maureen Flanagan, and others, conservation emerges not as a cohesive movement dominated by a few personalities, but as a constellation of continually renegotiated ideas. See Armitage, *Nature Study Movement*, 113; Maureen A. Flanagan, *America Reformed: Progressives and Progressivisms, 1890s-1920s* (New York: Oxford University Press, 2007), 176; Dorceta E. Taylor, *The Rise of the American Conservation Movement: Power, Privilege, and Environmental Protection* (Durham, NC: Duke University Press, 2016); Benjamin H. Johnson, *Escaping the Dark, Gray City: Fear and Hope in Progressive-Era Conservation* (New Haven, CT: Yale University Press, 2017), 7–8, 100. Johnson describes the complex relation between progressivism and conservation. As Johnson writes, "not all conservationists were Progressives, and not all Progressives were conservationists, but conservation was a central part of the Progressive quest to humanize industrial capitalism" (7).

27. For a discussion of the importance of hydrology in progressive technocracy, see Ellen Stroud, *Nature Next Door: Cities and Trees in the American Northeast* (Seattle: University of Washington Press, 2012), 25, 70–71. The often-tragic story of water resource management in the West has been chronicled by Donald Pisani and Donald Worster. See Donald J. Pisani, *Water and American Government: The Reclamation Bureau, National Water Policy, and the West, 1902–1935* (Berkeley: University of California Press, 2002); Donald Worster, *Rivers of Empire: Water, Aridity, and the Growth of the American West* (New York: Pantheon Books, 1985). Not all technocrats who studied hydrology were dedicated to "empire" building, however. Paul Sutter offers a far more generous interpretation of progressive forester Benton MacKaye in "A Retreat from Profit: Colonization, the Appalachian Trail, and the Social Roots of Benton MacKaye's Wilderness Advocacy," *Environmental History* 4 (October 1999): 553–77. See also Mark Fiege, *Irrigated Eden: The Making of an Agricultural Landscape in the American West* (Seattle: University of Washington Press, 1999).

28. "Friends of the Forests to Meet in Denver," *Denver Republican*, August 4, 1901. Forest History Society—US Forest Service Newspaper Clippings File, box 3 (American Forestry Association 4). Although the "Friends of the Forests" group—presumably shorthand for the American Forestry Association—was not part of the government, its interests and membership overlapped with those of the Forest Service, as evinced by the expansive collection of AFA documents in the USFS archives.

29. For schism and consensus within progressive conservation, see Johnson, *Escaping the Dark, Gray City*, especially 142–63; Taylor, *Rise of the American Conservation Movement*, 27.

30. Matthew Klingle, *Emerald City: An Environmental History of Seattle* (New Haven, CT: Yale University Press, 2007), 117, 66.

31. *Northwestern Horticulturist* (Minneapolis, MN), August 15, 1901. F. H. Newell spoke at a meeting of the Nebraska State Horticultural Society held in Kearney. Newell advanced a similar argument in his report "The Reclamation of the West," *Annual Report of the Board of Regents of the Smithsonian Institution . . . for the Year Ending June 30, 1903*

(Washington, DC: GPO, 1904). In the 1903 report Newell referred to climate improvement theories as a "popular delusion" that had "ensnared many emigrants" to the Great Plains (833–34). For more biographical information about Newell, see Tyrrell, *Crisis of the Wasteful Nation*, 101–10.

32. Although Marsh wrote that "it does not seem probable that the forests sensibly affect the total quantity of precipitation," he believed that trees influenced other climatic and hydrologic conditions. See George Perkins Marsh, *Man and Nature, or Physical Geography as Modified by Human Action* (New York: Scribner, 1864), 178.

33. *Eureka Californian*, August 19, 1901.

34. *Boston Evening Transcript*, July 29, 1901. It is unclear exactly which experiments the *Transcript* is alluding to. Several climate theorists—some associated with government department and some not—carried out experiments intended to measure the effect of forests on local climatic conditions. See, for example, L. C. Corbett [of West Virginia University], "Influence of Groves on the Moisture Content of the Air," *Forester* 3, no. 4 (April 1, 1897)—Forest History Society Archive—American Forestry Association Collection, box 7. Corbett set up "observatories" with "chemical thermometers" and "hygrometers" (for both wet bulb and dry bulb temperatures) in an attempt to determine the possible role of forests in shaping local climatic conditions. He found the results erratic and inconclusive, but in general "in favor of the forest station." He hoped that his admittedly incomplete results would spur further research and help demonstrate the utility of both afforestation in the West and forest conservation in the East: "It is my desire . . . to continue this work in the virgin forests of the east for the purpose of showing the comparative value of groves upon the prairie as well as to determine the relative influence of young and old forest covers upon soil and atmospheric conditions" (48).

35. Wilson was cited by the *Provo City (UT) Enquirer* on July 30, 1901. Several other newspapers mention his remarks, including the *Quincy (IL) Whig* of July 27, 1901, and the *Pittsburgh Dispatch* of July 30, 1901.

36. *Dallas News*, July 31, 1901. In later years, Pinchot would make more specific and authoritative claims about forests' influence on climate. In a 1909 report Hiram Chittenden of the US Army Corps of Engineers cited Pinchot as attributing a 10 percent increase in rainfall to the presence of forests. See Hiram H. Chittenden, "Forests and Reservoirs in Their Relation to Stream Flow with Particular Reference to Navigable Rivers," *Congressional Record* 43 (February 9, 1909): 945.

37. *Manchester (NH) American*, August 16, 1901.

38. See, for example, Royal S. Kellogg [forest agent, Bureau of Forestry], "Forest Planting in Western Kansas," US Department of Agriculture, Bureau of Forestry, Bulletin 52 (Washington, DC: GPO, 1904). Kellogg's conclusions were decidedly tentative, but he appeared more skeptical about human-induced climate change than his supervisor Gifford Pinchot.

39. *Forest Influences*, US Department of Agriculture, Forestry Division, Bulletin 7 (Washington, DC: GPO, 1893). Fernow's portion of the report also addressed one of the contentious subtopics within the climate debate: the question whether there existed an ideal ratio of forested areas to cleared areas. He argued that "the attempts to fix a certain percentage of forest cover needed for favorable climatic conditions of a country are devoid of all rational basis" (11). The complex and uncertain question of "forest influences," Fernow believed, could not be reduced to a single statistic.

40. *Youth's Companion*, July 19, 1894. The *Youth's Companion* was also quoted by the

Monthly Review of the Iowa Weather and Crop Service, Co-operating with the United States Signal Service 5, no. 9 (September 1894): 6–7; State Historical Society of Iowa, Iowa City, call no. QC.984.I6. That a scientific report like the Iowa *Monthly Review* cited the *Youth's Companion* (which in turn cited a government report) indicates that progressive conservation did not just trickle down from the realm of experts into popular culture. Rather, it developed as a dialectic between popular notions and scientific, technocratic theories.

41. Armitage, *Nature Study Movement*, 2.

42. Several scholars have challenged the interpretation of progressive conservation as a top-down movement shaped solely by experts. See Robert D. Johnston, *The Radical Middle Class: Populist Democracy and the Question of Capitalism in Progressive Era Portland, Oregon* (Princeton, NJ: Princeton University Press, 2003); Johnson, *Escaping the Dark, Gray City*, 146; Michael McGerr, *A Fierce Discontent: The Rise and Fall of the Progressive Movement in the United States, 1870–1920* (New York: Free Press, 2003), 246–47.

43. Joseph Cullon, "Legacies and Limitations: Environmental Historians Reconsider Progressive Conservation," *Journal of the Gilded Age and Progressive Era* 1, no. 2 (April 2002): 180.

44. George W. Rafter, "The Relation of Rainfall to Run-off," Water Supply and Irrigation Paper 80, United States Geological Survey, Department of the Interior (Washington, DC: GPO, 1903) 9.

45. Rafter, "Relation of Rainfall," 53. Rafter is referring to Hough's "Report upon Forestry" (US Department of Agriculture, 1877).

46. See Sutter, "Retreat from Profit," 555. See also Gordon B. Dodds, "The Stream-Flow Controversy: A Conservation Turning Point," *Journal of American History* 56 (June 1969), and Donald Pisani, "Forests and Reclamation, 1891–1911," in his *Water, Land, and Law in the West: The Limits of Public Policy, 1850–1920* (Lawrence: University of Kansas Press, 1996), 141–58.

47. Dodds, "Stream-Flow Controversy," 62.

48. Dodds, "Stream-Flow Controversy," 62–64.

49. Klingle, *Emerald City*, 69–71.

50. Chittenden, "Forests and Reservoirs," 925 for Chittenden's "sympathies," 958–59 for his discussion of deforestation.

51. Chittenden, "Forests and Reservoirs," 925–26. Chittenden also attempted to derive conclusions from his admittedly scanty data showing river stages for the Mississippi, Ohio, Tennessee, Missouri, and Connecticut Rivers. As Dodds has pointed out in his study of the Weeks Act, Chittenden lacked data showing the amount of logging that had taken place in each river's watershed. Without this information, his own claims remained as tentative as those of his foes. See Chittenden, 958–65, and Dodds, "Stream-Flow Controversy," 63.

52. Sutter, "Retreat from Profit," 572n8.

53. M. O. Leighton, A. C. Spencer, and B. MacKaye, *The Relation of Forests to Stream Flow* (Washington, DC: US Department of the Interior, Geological Survey, 1913), 2. Copy of report obtained from Forest History Society Archives.

54. MacKaye, "Forest Cover and Topography," in Leighton, Spencer, and MacKaye, *Relation of Forests to Stream Flow*, 38.

55. For a description of science as the guiding spirit of conservationism, see Ramachandra Guha, *Environmentalism: A Global History* (New York: Longman, 2000), 27.

56. See Chittenden, "Forests and Reservoirs," 942. For the characterization of distur-

bance as one of the core beliefs of Progressive Era conservation, see Cullon, "Legacies and Limitations," 180.

57. Sutter, "Retreat from Profit," 555, 556 for "community settlement."

58. Armitage, *Nature Study Movement*, 109. Armitage also writes that "modern science helped produce a culture based on bureaucracy, methodical thinking, adherence to system, and quantification; in short, modern science helped create the instrumental rationality that worried Max Weber as well as American nature lovers" (40–41).

59. *Hubbard (IA) Monitor*, August 2, 1901.

60. As Char Miller explains, Pinchot adopted Jeremy Bentham's aphorism about the "greatest good" and added the phrase "in the long run" to highlight foresters' need to consider the long-lasting influence of their work. See Miller, *Gifford Pinchot and the Making of Modern Environmentalism* (Washington, DC: Island Press, Shearwater Books, 2001), 155.

61. This mixture of old and new environmental rhetoric underscores Benjamin H. Johnson's characterization of Progressive Era conservation as both a break from earlier developments and as a series of "adaptations" drawn from "diverse traditions of conservationist thinking." See *Escaping the Dark, Gray City*, 8.

62. *Jacksonville (FL) Times Union*, February 6, 1899.

63. *Nebraska City (NE) Conservative* 3, no. 40 (April 11, 1901), FHS—USFS Newspaper Clipping File, box 4, Arbor Day Files.

64. For a quantitative study discussing the effects of intermittent droughts in the late 1880s and early 1890s on homesteading in the Great Plains, see Libecap and Hansen, "Rain Follows the Plow and Dryfarming Doctrine," 90–94.

65. Martin Mohler, "A New Departure in Agriculture," *Kansas Farmer*, November 29, 1893.

66. Mohler, "New Departure."

67. Alan Trachtenberg called Turner's frontier thesis "as much an invention of cultural belief as a genuine historical fact." Trachtenberg, *The Incorporation of America: Culture and Society in the Gilded Age* (New York: Hill and Wang, 1982), 14–19. Patricia Limerick famously articulated a more forceful critique, arguing that Turner overlooked the victims of expansionism and depicted the West only from the perspective of the East (and not as a place in itself). Patricia Nelson Limerick, *The Legacy of Conquest: The Unbroken Past of the American West* (New York: Norton, 1987).

68. Judge S. Emery, "Our Arid Lands," *Arena* 17, no. 3 (February 1897): 389–94. For more on irrigation boosterism and the belief that irrigation would unlock the latent potential of western landscapes, see David Nye, *America as Second Creation*, 215–30.

69. See Webb, *Great Plains*, 336–40.

70. *Topeka (KS) Daily Capital*, September 20, 1894.

71. *Topeka (KS) Daily Capital*, July 6, 1894.

72. Libecap and Hansen, "Rain Follows the Plow and Dryfarming Doctrine," 97. Private interests also supported the dry farming revolution: Great Plains railroads trumpeted the potential of dry farming and saw scientific soil culture as a "solution to the problem of settling their lands" (Libecap and Hansen, 101). For more on dry farming see Webb, *Great Plains*, 366, and Gilbert Fite, *The Farmer's Frontier, 1865–1900* (New York: Holt, Rinehart, and Winston, 1966). For agricultural experiment stations, see Jeremy Vetter, *Field Life: Science in the American West during the Railroad Era* (Pittsburgh, PA: University of Pittsburgh Press, 2016).

73. See David J. Wishart, *The Last Days of the Rainbelt* (Lincoln: University of Ne-

braska Press, 2013). US Census figures attest to the dramatic slowdown in settlement across Kansas and Nebraska during the 1890s. From 1870 to 1880, Kansas's population grew by 173 percent (from 364,339 to 996,096). Growth rates during the 1880s were less astronomical but remained very high, with an increase of 43 percent to a total of 1,428,108 by 1890. Kansas's decennial growth rate dropped precipitously during the 1890s, falling to 3 percent (1,470,495) before recovering somewhat during the first decade of the 1900s (to 15 percent). See Kansas Statistical Abstract—Enhanced Online Edition, http:// www.ipsr.ku.edu/ksdata/ksah/population/2pop1.pdf (accessed September 16, 2017). Nebraska experienced an even starker decline in decennial growth rate, going from 268 percent in the 1870s, to 134 percent in the 1880s, to a meager 0.3 percent in the 1890s. See http://population.us/ne/ (accessed September 16, 2017). Many rural counties in the western portions of both states experienced an even greater decline in growth, with some seeing marked demographic declines during the 1890s. See, for example, the 1890s population loss in Logan, Gove, and Wichita Counties on this database: http://ipsr.ku .edu/ksdata/ksah/population/2pop16.pdf (accessed September 16, 2017). Some late nineteenth- and early twentieth-century inhabitants of the Plains states viewed the decennial census as a crucial indicator of the welfare of their states. Contention ensued after the 1900 census showed a decline in Omaha's population. Some alleged that the census total for 1900 was too low, but other Nebraskans admitted that the 1890 figure had likely been padded, setting the stage for a disappointment at the end of the century. See *Omaha Daily News*, August 23, 1900, and the Nebraska State Historical Society's brief piece "Census of 1900," http://www.nebraskahistory.org/publish/publicat/timeline/ census_of_1900.htm (accessed September 16, 2017).

74. Libecap and Hansen, "Rain Follows the Plow and Dryfarming Doctrine," 113–14.

75. See Mary Hargreaves, *Dry Farming in the Northern Great Plains: 1900–1925* (Cambridge, MA: Harvard University Press, 1957), 546; Libecap and Hansen show that farm failure remained frequent into the 1910s and 1920s; see Libecap and Hansen, "Rain Follows the Plow and Dryfarming Doctrine," 113.

76. *Wichita (KS) Eagle*, January 18, 1901. Kansans might have had the last laugh, however, since Henry died penniless in 1914 after the failure of his many efforts to support irrigation in Colorado. See the Theodore C. Henry entry in the Kansas Historical Society's online "Kansapedia," https://www.kshs.org/kansapedia/theodore-c-henry/18164 (accessed September 16, 2017).

77. Newell, "Reclamation of the West," 833–34.

78. T. E. Haines discussed "pond theory" in his piece "What Has Tile Drainage Done for Iowa?," *Monthly Review of the Iowa Weather and Crop Service, Co-operating with the United States Signal Service* 2, no. 3 (March 1891): 7–8.

79. John Wesley Powell, "Trees on Arid Lands," *Science* 12, no. 297 (October 12, 1888).

80. *Topeka (KS) Daily Capital*, September 10, 1890.

81. J. R. Sage, "Influence of Forests on Climate in Iowa," November 23, 1893, speech before the Iowa Horticultural Society, reproduced in the "Current Notes" section of the *American Meteorological Journal* 10, no. 14 (March 1894): 478.

82. *Annual Report of the Iowa Weather and Crop Service in Co-operation with the U.S. Department of Agriculture, Weather Bureau, for the Meteorological Year 1890* (Des Moines, IA: G. H. Ragsdale, 1891), 7.

83. In 1892 the *Iowa Farmer and Breeder* criticized a "contemporary farm journal" for claiming that if tile drainage continued Iowa would become a "genuine desert, and people will look upon a small pond of water with joy." The unspecified farm journal

issued a warning to anyone undertaking drainage initiatives: "Leave our ponds alone or the farmers will soon be compelled to petition the [legislature] to irrigate the arid lands in Iowa." *Farmer and Breeder* retorted by claiming that if all the boggy areas between the Mississippi River and the Rocky Mountains were "completely drained, the average rainfall would be increased and more easily distributed." The exchange between the two publications was cited by the *Monthly Review of the Iowa Weather and Crop Service* 3, no. 4 (April 1892): 4–5.

84. J. R. Sage, "The Practical Value of Reliable Crop and Weather Reports," *Monthly Review of the Iowa Weather and Crop Service, Co-operating with the United States Signal Service* 1, no. 8 (November 1890): 3; Des Moines, IA: R. H. Ragsdale. State Historical Society of Iowa, Iowa City—Call No. QC.984.I6. Sage proved to be an enthusiastic, if highly selective, proponent of expansionism. He described Iowa as a "vast agricultural empire, unequalled in respect to fertility of soil and salubrity by climate by any body of land of like area on the face of the globe." But he voiced his doubts about continued expansion into other portions of the West: "Nothing can fully compensate for the lack of rainfall in the growing season, for only a small portion of any arid region can be reclaimed by irrigation." See *Annual Report of the Iowa Weather and Crop Service* for 1890, 9–10.

85. Sage, "Practical Value," 3.

86. William Hammond Hall, "Influence of Parks and Pleasure Grounds," *Biennial Report of the Engineer of the Golden Gate Park, for Term Ending Nov 30th, 1873.*

87. Isaac Monroe Cline, *Summer Hot Winds on the Great Plains*, paper read before the Philosophical Society of Washington, January 20, 1894 (Washington, DC: Philosophical Society, 1894), 348. E. F. Stephens, "What Has the Timber Claim Law Done for Nebraska?," *Annual Report of the Nebraska State Horticultural Society for the Year 1897 Containing the Proceedings of the Annual Meeting Held at Lincoln, January, 1897* (Lincoln, NE: State of Nebraska, 1897), 52. H. C. Price, "Forestry and Its Effect on Western Climate," *Proceedings of the Iowa Park and Forestry Association. Second Annual Meeting, Des Moines, Iowa, December 8, 9, 10, 1902* (Iowa City: Iowa Park and Forestry Association, 1903), 31–33.

88. Robert Gardner, "Constructing a Technological Forest: Nature, Culture, and Tree-Planting in the Nebraska Sand Hills," *Environmental History* 14, no. 2 (April 2009): 280, 288. Joel Jason Orth, "The Conservation Landscape: Trees and Nature on the Great Plains" (PhD diss., Iowa State University, 2004). Joel Orth, "Directing Nature's Creative Forces: Climate Change, Afforestation, and the Nebraska National Forest," *Western Historical Quarterly* 42, no. 2 (Summer 2011): 197–217.

89. Orth, "Conservation Landscape," 140. See also Joel Orth, "The Shelterbelt Project: Cooperative Conservation in 1930s America," *Agricultural History* 81, no. 3 (Summer 2007): 333–57.

90. Orth, "Conservation Landscape," 145.

91. See Orth, "Conservation Landscape," 141–46. For more on the contentious debates surrounding the shelterbelt program, see Wilmon H. Droze, *A History of Tree Planting in the Plains States* (Denton: Texas Woman's University, 1977); David Moon, *The American Steppes: The Unexpected Russian Roots of Great Plains Agriculture, 1870s-1930s* (Cambridge: Cambridge University Press, 2020), 343–45.

92. Moon, *American Steppes*, 309.

93. Raphael Zon, *Forests and Water in Light of Scientific Investigation* (Washington, DC: GPO, 1927), 1 and 17.

94. Raphael Zon, "How the Forests Feed the Clouds," in *Science Remaking the World* (New York: Doubleday, Page, 1923), 215–21. For a later example of Zon's views on climate and forestry, see *Possibilities of Shelterbelt Planting in the Plains Region,* prepared under the direction of the Lake States Forest Experiment Station, US Forest Service (Washington, DC: GPO, 1935).

95. Neil Maher, *Nature's New Deal: The Civilian Conservation Corps and the Roots of the American Conservation Movement* (Oxford: Oxford University Press, 2008), 11, 214–24.

96. The *Iowa Farmer and Breeder* was cited by the *Monthly Review of the Iowa Weather and Crop Service, Co-operating with the United States Signal Service* 3, no. 4 (April 1892): 4–5.

Conclusion

1. See Ian Tyrrell, *Crisis of the Wasteful Nation: Empire and Conservation in Theodore Roosevelt's America* (Chicago: University of Chicago Press, 2015), 12, 18; Andrew Offenburger, *Frontiers in the Gilded Age: Adventure, Capitalism, and Dispossession from Southern Africa to the U.S.-Mexican Borderlands, 1880–1917* (New Haven, CT: Yale University Press, 2019).

2. M. O. Baldwin, "The Panama Canal and Its Possible Influences upon the Ocean Currents and upon Climate," *Kansas City Review of Science and Industry,* vol. 8 (Kansas City, MO: Ramsey, Millet, and Hudson, 1885), 447. As James Rodger Fleming has shown, the notion of changing large-scale climate patterns by constructing a canal across the Isthmus of Panama dates back to Thomas Jefferson. See Fleming, *Historical Perspectives on Climate Change* (Oxford: Oxford University Press, 1998), 29–30.

3. Marcus Hall, *Earth Repair: A Transatlantic History of Environmental Restoration* (Charlottesville: University of Virginia Press, 2005). As Hall shows, the life and work of George Perkins Marsh epitomizes the transatlantic exchange of environmental ideas.

4. Franklin B. Hough, *Essay on the Climate of the State of New York* (Albany, NY: Van Benthuysen, 1857), 43–44. For other examples of Euro-American climate theorists' drawing inspiration from Dove, see Lorin Blodget, *Climatology of the United States* (Philadelphia: J. B. Lippincott, 1857), 406; John H. Klippart, "Forests: Their Influence upon Soil and Climate," *Fifteenth Annual Report of the Ohio State Board of Agriculture, for the Year 1860* (Columbus, OH: Richard Nevins, 1861), 265–69.

5. Franklin B. Hough, *Report upon Forestry* (Washington, DC: GPO, 1878), 230–35; Cleveland Abbe, *A First Report on the Relations between Climates and Crops* (Washington, DC: GPO, 1905), 7–8; Raphael Zon, *Forests and Water in Light of Scientific Investigation* (Washington, DC: GPO, 1927), 3–5.

6. For France, see George F. Swain, "The Influence of Forests upon the Rainfall and upon the Flow of Forests," *American Meteorological Journal* 7, no. 5 (November 1888): 297. For Italy, see F. B. Hough, Speech for the American Association for the Advancement of Science, box 39, folder 1, Hough Papers, New York State Library, 25–28. For British colonial forestry, see A. W. Greely, *American Weather* (New York: Dodd, Mead, 1888), 155–57.

7. Woiekoff's name is sometimes transliterated as Voeikov. For Americans citing Woiekoff, see Greely, *American Weather,* 157. Woiekoff's assessment of Curtis's rainfall theories is from an October 6, 1893, letter from Woiekoff to the *American Meteorological Journal.* See A. Woiekoff, "Causes of Rainfall and Surface Conditions," *American Meteorological Journal* 10, no. 9 (January 1894).

8. Anya Zilberstein, *A Temperate Empire: Making Climate Change in Early America* (New York: Oxford University Press, 2017); Fabien Locher and Jean-Baptiste Fressoz, "Modernity's Frail Climate: A Climate History of Environmental Reflexivity," *Critical Inquiry* 38, no. 3 (Spring 2012): 579–98.

9. For a comparative analysis of environmental theories and colonial settlement projects in arid and semiarid landscapes, see Diana K. Davis, *The Arid Lands: History, Power, Knowledge* (Cambridge, MA: MIT Press, 2016). See also David Moon, *The American Steppes: The Unexpected Russian Roots of Great Plains Agriculture, 1870–1930* (Cambridge: Cambridge University Press, 2020).

10. Jonathan Levy, *Freaks of Fortune: The Emerging World of Capitalism and Risk in America* (Cambridge, MA: Harvard University Press, 2012), 14. Levy traces the origins of the concept of "radical uncertainty" to early twentieth-century economic theorists. See Jonathan Levy, "Radical Uncertainty," *Critical Quarterly* 62, no. 1 (April 2020): 15–28.

11. Joshua Howe's work on recent climate science has revealed the limits of the "forcing function" of scientific knowledge. See Howe, *Behind the Curve: Science and the Politics of Global Warming* (Seattle: University of Washington Press, 2014). Rachel Rothschild's history of the acid rain controversy discusses the "precautionary principle," a "rationale for acting in the face of uncertainty." See Rothschild, *Poisonous Skies: Acid Rain and the Globalization of Pollution* (Chicago: University of Chicago Press, 2019), 151.

12. Mike Hulme, *Why We Disagree about Climate Change* (Cambridge: Cambridge University Press, 2009), 82.

13. Steven Yearley has written that the "revocable and provisional" character of scientific facts sometimes "fails to deliver the decisiveness and moral certainty" environmentalists seek. See Yearley, "The Environmental Challenge to Science Studies," in *Handbook of Science and Technology Studies* (Thousand Oaks, CA: Sage, 1995), 461–63.

BIBLIOGRAPHY

Archives

American Geographical Society Library
American Philosophical Society Library
Cornell University Archives—Bernhard Fernow Papers
Forest History Society—George H. Wirt Collection
Kansas Historical Society
Library of Congress—Adolphus Greely Papers
Missouri History Museum—Richard Smith Elliott Papers
National Archives, College Park, Maryland—"Records Relating to the Rain Producing
    Experiment, 1892"
Nebraska State Historical Society
New York State Library—Franklin B. Hough Papers
Peabody-Essex Museum—George Loring Papers
State Historical Society of Iowa
University of Iowa Archives—Gustavus Hinrichs Papers
University of Kansas Archives, Spencer Library
University of Nebraska Special Collections

Newspapers and Periodicals

American Meteorological Journal
Arena
Boston Evening Transcript
Christian Examiner
Dallas News
Denver Republican
Dodge City (KS) Times
Eureka Californian
Garden City (KS) Herald
Hays City (KS) Sentinel

Hubbard (IA) Monitor
Iowa Farmer and Breeder
Jacksonville (FL) Times Union
Junction City (KS) Union
Kansas City (MO) Journal
Kansas City (MO) Star
Kansas Farmer
Larned (KS) Chronoscope
Larned (KS) Press
Leavenworth (KS) Times
Manchester (NH) American
Nation
Nebraska City (NE) Conservative
Nebraska Farmer
New York Times
North American Review
Northwestern Horticulturalist (Minneapolis)
Omaha (NE) Daily News
Pacific Rural Press
Phillips County (KS) Inter Ocean
Pittsburgh Dispatch
Prairie Farmer
Provo (UT) Enquirer
Quincy (IL) Whig
Science
Scientific American
Southern Lumberman
Topeka (KS) Commonwealth
Topeka (KS) Daily Capital
Topeka (KS) Daily Herald
Topeka (KS) Weekly Tribune
Virginia City (NV) Enterprise
Wichita (KS) Eagle
Youth's Companion

Primary Sources

Abbe, Cleveland. *A First Report on the Relations between Climates and Crops*. Washington, DC: GPO, 1905.

Adams, Charles Francis. "The Rainfall on the Plains." *Nation*, November 14, 1887.

Allan, J. T. *Nebraska and Its Settlers: What They Have Done and How They Do It; Its Crops and People*. Omaha, NE: Union Pacific Company Land Department, 1883.

Allen, Martin. "Brief Historical Sketch of Ellis County up to the Close of the Centennial Year." *Hays City (KS) Sentinel*, January 18, 1878.

Andrews, L. F. "The State of Iowa: Its Topography, Drainage, Fertility, Climate, and Healthfulness." *Monthly Review of the Iowa Weather and Crop Service, Co-operating with the United States Signal Service*, April and May 1890. State Historical Society of Iowa, Iowa City—Call No. QC.984.I6 Des Moines, IA: R. H. Ragsdale.

Aughey, Samuel. "The Increasing Need for Forests." *Nebraska Farmer* 1, no. 10 (October 1877).

———. *The Physical Geography and Geology of Nebraska.* Omaha, NE: Daily Republican Book and Job Office, 1880.

Baker, F[loyd] P[erry]. *Preliminary Report on the Forestry of the Mississippi Valley and Tree Planting on the Plains.* Washington, DC: GPO, 1883.

Baldwin, M. O. "The Panama Canal and Its Possible Influences upon the Ocean Currents and upon Climate." *Kansas City Review of Science and Industry,* vol. 8. Kansas City, MO: Ramsey, Millet, and Hudson, 1885.

Beadle, J. H. *The Undeveloped West, or Five Years in the Territories.* Philadelphia: National Publishing Company, 1873.

Bent, Silas. "Meteorology of the Mountains and Plains of North America, as Affecting the Cattle-Growing Industries of the United States." *American Meteorological Journal* 1, no. 11 (March 1885).

Bidwell, W. D. "The Relation of Climatic Changes to Disease." *Kansas Medical Journal* 1, no. 7 (November 1889).

Blodget, Lorin. *Climatology of the United States.* Philadelphia: J. B. Lippincott, 1857.

———. "Forest Cultivation on the Plains." *Report of the Commissioner of Agriculture for the Year 1872.* Washington, DC: GPO, 1872.

Bowles, Samuel. *Across the Continent: A Summer's Journey to the Rocky Mountains, the Mormons, and the Pacific States.* Springfield, MA: Samuel Bowles, 1865.

Boynton, C. B., and T. B. Mason. *A Journey through Kansas; with Sketches of Nebraska: Describing the Country, Climate, Soil, Mineral, Manufacturing, and Other Resources.* Cincinnati, OH: Moore, Wilstach, Keys, 1855.

Brevoort, Elias. *New Mexico: Her Natural Resources and Attractions.* Santa Fe, NM: Elias Brevoort, 1874.

Brockett, L. P. *Our Western Empire, or The New West beyond the Mississippi.* Columbus: William Garretson, 1881.

Budd, J. L. [Iowa Board of Forestry]. "Possible Modification of Our Prairie Climate." In Preliminary Newspaper Report, *Sixth Annual Meeting of the American Forestry Congress,* Held in Springfield, IL, 1887. Springfield, IL: State Register Book and Job Print, 1887.

Burton, Richard F. *The City of the Saints: Across the Rocky Mountains to California.* 1861, 1881. Reprint, New York: Knopf, 1963.

Clarke, F. M. "Shall We Build Reservoirs?" *Annual Report of the Colorado State Board of Horticulture and State Agricultural and Forestry Association* (1887–88).

Chittenden, Hiram H. "Forests and Reservoirs in Their Relation to Stream Flow with Particular Reference to Navigable Rivers." *Congressional Record* 43 (February 9, 1909).

Cleveland, H. W. S. *Landscape Architecture as Applied to the Wants of the West.* 1873. Reprint, Amherst: University of Massachusetts Press, 2002.

Cline, Isaac Monroe. "Summer Hot Winds on the Great Plains." Paper read before the Philosophical Society of Washington, January 20, 1894. Washington, DC: Philosophical Society of Washington, 1894.

Cooper, Ellwood. *Forest Culture and Eucalyptus Trees.* San Francisco, CA: Cubery, 1876.

Corbett, L. C. "Influence of Groves on the Moisture Content of the Air." *Forester* 3, no. 4 (April 1, 1897).

Cronk, Corydon P. "Influence of Forests on Climate and Agriculture." *Maryland State*

Weather Service Monthly Report 3, no. 6. Washington, DC: US Department of Agriculture, Weather Bureau, October 1893.

Culbertson, Harvey. "Meteorology." *Annual Report of the Nebraska State Horticultural Society, 1885.* Lincoln, NE: State Journal Company, 1887.

Curley, Edwin A. *Nebraska 1875: Its Advantages, Resources, and Drawbacks.* New York: American News Company, 1875.

Curtis, George E. "Analysis of the Causes of Rainfall with Special Relation to Surface Conditions." *American Meteorological Journal* 10, no. 6 (October 1893).

———. "The Facts about Rain-Making." "Records Relating to the Rain-Producing Experiment, 1892." *Engineering Magazine,* National Archives, College Park, RG 95, entry 123, box 1, 540.

———. "The Facts about Rain-Making." *American Meteorological Journal* 9, no. 6 (October 1892).

———. "Review of Rainfall Laws, Deduced from Twenty Years' Observation." *American Meteorological Journal* 10 (April 1894).

———. "The Trans-Mississippi Rainfall Problem Restated: The Rainfall in Its Relation to Kansas Farming." *American Meteorological Journal* 5, no. 2 (June 1888).

Cutting, Hiram Adolphus. *Forests of Vermont.* Montpelier: Vermont Watchman and State Journal Press, 1886.

———. *Lectures on Plants, Fertilization, Insects, Forestry, Farm Homes, Etc.* Montpelier, VT: Foreman Steam Printing, 1882.

Davis, William Morris. "The Theories of Artificial and Natural Rainfall." *American Meteorological Journal* 8, no. 1 (March 1892).

Dillon, Sidney [President of the Union Pacific Railway Company]. "The West and the Railroads." *North American Review* 152, no. 413 (April 1891).

Disturnell, John. *Influence of Climate.* New York: D. Van Nostrand, 1867.

Dodds, S. W. "What Causes the Cyclones." *Proceedings of the American Forestry Association at the Summer Meeting, Held in Quebec, September 2–5, 1890, and at the Ninth Annual Meeting, Held at Washington, December 30, 1890* (Washington DC, 1891).

Dorsey, Stephen. "Land Stealing in Mexico, a Rejoinder." *North American Review* 145 (October 1887).

Dyrenforth, Robert. "Can We Make It Rain?" *North American Review* 153, no. 419 (October 1891).

Egleston, N. H. *Arbor Day Leaves: A Complete Programme for Arbor Day Observance, Including Readings, Recitation, Music, and General Information.* New York: American Book Company, 1893.

Elliott, Richard Smith. "Climate of Kansas." Submitted to Pond Creek Station, Kansas Pacific Railway, September 22, 1870. *Annual Report of the Board of Regents of the Smithsonian Institution, Showing the Operations, Expenditures, and Condition of the Institution for the Year 1870.* H.R., 42nd Cong., 1st Sess. Ex. Doc. No. 20. Washington, DC: GPO, 1871.

———. *Notes Taken in Sixty Years.* St. Louis, MO: R. P. Studley, 1883.

Emery, Judge S. "Our Arid Lands." *Arena* 17, no. 3 (February 1897).

Ensign, Edgar T. "Report on the Forest Conditions of the Rocky Mountains." Department of Agriculture, Forestry Division, Bulletin 2, *Report on the Forest Conditions of the Rocky Mountains, and Other Papers.* Washington, DC: GPO, 1888.

Fernow, Bernhard E. "Introduction and Summary of Conclusions." US Department of Agriculture, Forestry Division, Bulletin 7, *Forest Influences.* Washington, DC: GPO, 1893.

Finley, John P. *Certain Climatic Features of the Two Dakotas*. Washington, DC: GPO, 1893.

Forest Influences. US Department of Agriculture, Forestry Division, Bulletin 7. Washington, DC: GPO, 1893.

Foster, J. W. *The Mississippi Valley*. Chicago: S. C. Griggs, 1869.

Furnas, Robert. *Arbor Day*. Lincoln, NE: State Journal Company, 1888.

Gale, Elbridge. "Forest Tree Culture." Paper read before the Kansas State Horticultural Society, Manhattan, December 7(?), 1878.

Gannett, Henry. "Do Forests Influence Rainfall?" *Science* 11, no. 257 (January 6, 1888).

Glassford, W. A. "Charts and Tables of Rainfall on the Pacific Slope." In *Report of Rainfall in Washington Territory, Oregon, California, Idaho, Nevada, Utah, Arizona, Colorado, Wyoming, New Mexico, Indian Territory, and Texas, for from Two to Forty Years*. US Signal Corps. H.R., 50th Cong., 1st Sess., Ex. Doc. No. 91. Washington, DC: Government Printing Office, 1889.

———. "Climate of Arizona, with Particular Reference to the Rainfall and Temperature, and Their Influence upon the Irrigation Problems of the Territory." In *Irrigation and Water Storage in the Arid Regions*, US Signal Office. Washington, DC: GPO, 1891.

———. "Climate of New Mexico." In "Report on the Climatology of the Arid Region of the United States, with Reference to Irrigation," by Adolphus Greely. Submitted as part of a *Report on the Climate of New Mexico* by the Chief Signal Officer of the Signal Corps. Washington, DC: GPO, 1891.

Gott, Benjamin. "Forest Tree-Planting: The Results and Advantages to Farmers." In *American Journal of Forestry*, edited by Franklin B. Hough, vol. 1 (September 1882-October 1883). Cincinnati, OH: Robert Clarke, 1882–83.

Greely, Adolphus. *American Weather*. New York: Dodd, Mead, 1888.

———. *Report of Rainfall in Washington Territory, Oregon, California, Idaho, Nevada, Utah, Arizona, Colorado, Wyoming, New Mexico, Indian Territory, and Texas, for from Two to Forty Years*. H.R., 50th Cong., 1st Sess., Ex. Doc. No. 91. Washington, DC: GPO, 1889.

———. "Report on the Climatology of the Arid Region of the United States, with Reference to Irrigation." Submitted as part of a *Report on the Climate of New Mexico* by the Chief Signal Officer of the Signal Corps. Washington, DC: GPO, 1891.

———. "Some Peculiarities of the Rainfall of Texas." *Philosophical Society of Washington, Bulletin*, vol. 12. Washington, DC, April 1892.

Guild, Edgar. "Western Kansas: Its Geology, Climate, Natural History, Etc." *Kansas City Review of Science and Industry* 3 (December 1879).

Haines, T. E. "What Has Tile Drainage Done for Iowa?" *Monthly Review of the Iowa Weather and Crop Service, Co-operating with the United States Signal Service* 2, no. 3 (March 1891).

Hall, William Hammond. "Influence of Parks and Pleasure Grounds." *Biennial Report of the Engineer of the Golden Gate Park, for Term Ending November 30th, 1873*.

Hamilton, Patrick. *The Resources of Arizona*. San Francisco, CA: Bancroft, 1883.

Harrington, Mark W. "Climate of Santa Fe." *American Meteorological Journal* 2, no. 3 (July 1885).

———. "Is the Rain-Fall Increasing on the Plains?" *American Meteorological Journal* 4, no. 8 (December 1887).

———. "Weather-Making, Ancient and Modern." *Annual Report of the Board of the Regents of the Smithsonian Institution, 1894*. Washington, DC: GPO, 1896.

Hart, F. W. "Report on Forestry—Needs of Our Prairie State." *Transactions of the Iowa State Horticultural Society for 1879*. Des Moines, IA: F. M. Mills, 1880.

Hawn, Frederic. "Climatic Changes." *Western Homestead* (Leavenworth, KS) 3, no. 3 (August 1880).

———."Influence of Forests on Climate: Can the Plains Be Reclaimed by Tree-Planting?" *Kansas Magazine* 3, no. 6 (June 1873).

———. "Source of Rains in Kansas." *Quarterly Report of the Kansas State Board of Agriculture for the Quarter Ending September 30, 1881*. Topeka: Kansas Publishing House, 1881.

Hay, John. "Atmospheric Absorption and Its Effect upon Agriculture." *Proceedings of the Eighteenth Annual Meeting of the Kansas State Board of Agriculture*. Topeka, KS, 1890.

Hayden, Ferdinand V. *The First, Second, and Third Annual Reports of the U.S. Geological Survey of the Territories for the Years 1867, 1868, and 1869 under the Department of the Interior*. Washington, DC: GPO, 1873.

———. *The Great West*. Bloomington, IL: Charles R. Brodix, 1880.

———. *Preliminary Report of the United States Geological Survey of Wyoming and Portions of Contiguous Territories*. Washington, DC: GPO, 1871.

Hayes, W. G. M. "Professor Aughey Lived a Useful Life: Notes of the Address Delivered by Rev. W. G. M. Hayes at the Obsequies of the Late Professor Samuel Aughey" (1912). University of Nebraska Special Collections, RG 52//01, box 13.

Hazen, Henry Allen. "Variation of Rainfall West of the Mississippi River." Signal Service Notes, no. 7, US War Department. Washington, DC: Office of the Chief Signal Officer, 1883.

Hazen, W[illiam] B[abcock]. "The Great Middle Region of the United States, and Its Limited Space of Arable Land." *North American Review* 120, no. 246 (January 1875).

———. *Our Barren Lands: The Interior of the United States West of the One-Hundredth Meridian and East of the Sierra Nevada*. Cincinnati, OH: R. Clarke, 1875.

Henry Theodore C. Addresses on "Kansas Stock Interests" and "Kansas Forestry." Abilene, KS: Gazette Steam Printing Office, 1882.

Hilgard, E. W. "Climates of the Pacific Slope." *Report Made under the Direction of the Commissioner of Agriculture*, by E. W. Hilgard, T. C. Jones, and R. W. Furnas. Washington, DC: Government Printing Office, 1882. Morton Pamphlet Collection, vol. 1, Nebraska State Historical Society.

Hilton, H[ugh] R[ankin]. "Effects of Civilization on the Climate and Rain Supply of Kansas." Lecture delivered before the Scientific Club of Topeka, March 31, 1880. Spencer Library, University of Kansas Archives, RH C4318.

———. "Influence of Climate and Climatic Changes upon the Cattle Industry of the Plains." *Report for the Kansas State Board of Agriculture* (1888).

———. "Moisture Economy in Kansas." *American Meteorological Journal* 6, no. 5 (September 1889-October 1889).

Hinrichs, Gustavus. *Biennial and Annual Reports of the Iowa State Weather Service*. Des Moines: Iowa State Printer.

———. "Faith and Science." Lecture Delivered by Prof. Gustavus Hinrichs before the students of the [Iowa] State University. Undated [probably 1867]. University of Iowa Library, University Archives, G. Hinrichs Papers, box 1. RG 99.0039.

———. *Notes on Cloud Forms and the Climate of Iowa*. Central Station, Iowa Weather Service, 1883.

———. "Rainfall and Timber in Iowa." *Transactions of the Iowa State Horticultural Society for 1879*. Des Moines, IA: F. M. Mills, 1880.

———. *Rainfall Laws Deduced from Twenty Years of Observation.* Washington, DC: Weather Bureau, 1893.

———. "Tornadoes and Derechos." *American Meteorological Journal* 5, no. 9 (January 1889).

Hough, Franklin B. *The Elements of Forestry.* Cincinnati, OH: Robert Clarke, 1882.

———. *Essay on the Climate of the State of New York.* Albany, NY: Van Benthuysen, 1857.

———. "Letter from Dr. Franklin B. Hough, n Regard to the Effect of Forests in Increasing the Amount of Rainfall," app. 14. H.R., 48th Cong., 2nd Sess., Ex. Doc. Washington, DC: GPO, 1885.

———. *Report upon Forestry.* Washington, DC: Government Printing Office, 1878.

———. "The Value of American Timber Lands." In *The Proper Value and Management of Government Timber Lands and the Distribution of North American Forest Trees.* Department of Agriculture, Miscellaneous, Special Report no. 5. Washington, DC: GPO, 1884.

Hutchinson, Clinton C. *Resources of Kansas: Fifteen Years' Experience.* Topeka, KS: Author, 1871.

"The Influence of Forests on the Quantity and Frequency of Rainfall." *Science* 12, no. 303 (November 23, 1888).

Inman, Henry. "On Climatic Changes in the Prairie Region of the United States." *Kansas City Review of Science and Industry* 2, July 1878.

"Is Our Average Rainfall Diminishing?" *Scientific American* 54, no. 7 (February 13, 1886).

"Is the Climate of Kansas Changing?" *Larned (KS) Press,* August 8, 1878.

James, E. J. "The Government in Its Relation to the Forests." Department of Agriculture, Forestry Division, Bulletin 2. Washington, DC: GPO, 1888.

Kellogg, Royal S. "Forest Planting in Western Kansas." US Department of Agriculture, Bureau of Forestry, Bulletin 52. Washington, DC: GPO, 1904.

Kelsey, S. T. "Forest Trees and Hedges for the Western Prairies" (1872). Pamphlet held by the Kansas Historical Society (634.9 Pam. V. 2).

Kern, Maximilian G. "The Relation of Railroads to Forest Supplies and Forestry." Department of Agriculture, Forestry Division, Bulletin 1. "Report on the Relation of Railroads to Forest Supplies and Forestry." Washington, DC: GPO, 1887.

Kinney, Abbot. "Forests and Floods in Southern California." *American Meteorological Journal* 3, no. 9 (January 1887).

Klippart, John H. "Forests: Their Influence upon Soil and Climate." *Fifteenth Annual Report of the Ohio State Board of Agriculture for the Year 1860.* Columbus, OH: Richard Nevins, 1861.

Knox, M. V. B. "Climate and Brains." *Transactions of the Kansas Academy of Science,* vol. 5. Topeka, KS: George W. Martin, 1877.

Leighton, M. O., A. C. Spencer, and B. MacKaye. *The Relation of Forests to Stream Flow.* Washington, DC: US Department of the Interior, Geological Survey, 1913.

Logan, C. A. "Report on the Sanitary Conditions of the State of Kansas." 1865. University of Kansas, Spencer Library, RH C5773.

Loomis, Elias. "Contributions to Meteorology, Being Results Derived from an Examination of the Observations of the United States Signal Corps, and from Other Sources." *American Journal of Science and Arts,* 3rd ser., 13, no. 73 (January 1877).

Lotbinière, H. G. Joly de. "Tree Planting on the Prairies." *Proceedings of the American Forestry Association at the Summer Meeting, Held in Quebec, September 2–5, 1890, and* at the Ninth Annual Meeting, Held at Washington, December 30, 1890. Washington, DC, 1891.

Lovewell, J. T. "Human Agency in Changing or Modifying Climate." *Quarterly Report of the Kansas State Board of Agriculture for the Quarter Ending March 31, 1892.* Topeka, KS: Hamilton Printing Company, 1892.

———. "Kansas Weather Service." *Transactions of the Kansas Academy of Science for 1879–1880.* Topeka, KS: Geo. W. Martin, Kansas Publishing House, 1881.

———. "Kansas Meteorology." *Fourth Biennial Report of the State Board of Agriculture to the Legislature of the State of Kansas, for the Years 1883–1884.* Topeka: Kansas Publishing House: T. D. Thacher, 1885.

Loving, George B. *Letter from George B. Loving, Esq., of Fort Worth, Tex., in Regard to the Losses of Cattle during the Winter of 1884-'85, the Decline in the Value of Stock, and the Future of the Stock-Growing Interests of Texas.* H.R., 48th Cong., 2nd Sess., Ex. Doc., app. 15. Washington, DC: GPO, 1885.

Macomber, J. K. "Adaptability of Prairie Soils for Timber Growth." *Transactions of the Iowa State Horticultural Society for 1879.* Des Moines, IA: F. M. Mills, 1880.

Mahon, Charles, J. H. Renshawe, W. H. Graves, and H. Lindenkohl. *Map of Utah Territory, Representing the Extent of the Irrigable and Pasture Lands.* Department of the Interior. US Geographical and Geological Survey of the Rocky Mountain Region, 1878. Copy obtained from the American Geographical Society Library.

Marsh, George Perkins. *Man and Nature.* Edited by David Lowenthal. 1864. Reprint, Seattle: University of Washington Press, 2003.

McConnell, H. K. "Rainfalls of Kansas." *Osage County Chronicle,* March 30, 1882. Kansas Historical Society, Rain and Rainfall Clippings, 551.57R.

Meehan, Thomas. "Forests the Result and Not the Cause of Climate." *Prairie Farmer* 44, no. 47 (November 22, 1873).

Mohler, Martin. "Kansas Agriculture, Prospectively Considered." In *Report of the Kansas State Board of Agriculture for the Quarter Ending March 31, 1888.* Topeka: Kansas Publishing House, 1888.

———. "A New Departure in Agriculture." *Kansas Farmer,* November 29, 1893.

Moore, Willis L. "A Report on the Influence of Forests on Climate and on Floods." House of Representatives, United States Committee on Agriculture. Washington, DC: GPO, 1910.

Morton, J. Sterling. "Arbor Day." In *Proceedings of the American Forestry Congress* (1885).

———. "Arbor Day: Its Origin and Growth." Address of J. Sterling Morton Delivered April 22, 1886, at the State University, Lincoln, NE. Nebraska State Historical Society, J. Sterling Morton Pamphlet Collection, vol. 70.

Murphy, E[dward]. C[harles]. "Is the Rainfall in Kansas Increasing?" *Transactions of the Kansas Academy of Science,* vol. 13. Topeka, KS: Hamilton Printing Company, 1893.

Newell, F. H. "The Reclamation of the West." *Annual Report of the Board of Regents of the Smithsonian Institution . . . for the Year Ending June 30, 1903.* Washington, DC: GPO, 1904.

Nimmo, Joseph. "Report on the Internal Commerce of the United States." Department of the Treasury, Bureau of Statistics. Washington, DC: GPO, 1885.

Nipher, Francis E. "Report on Missouri Rainfall, with Averages for Ten Years Ending December 1887." *Transactions of the Academy of Science of St. Louis,* vol. 5, 1886–1891. St. Louis, MO: R. P. Studley, 1892.

Northrop, B. G. "Forests and Floods." *Report of Secretary of Connecticut Board of Agriculture.* Hartford, CT: Case, Lockwood, and Brainerd, 1885.

Noyes, Isaac P. "A New View of the Weather Question." *Kansas City Review of Science and Industry* 2 (July 1878).

"One Hundredth Meridian" (pamphlet). *Topeka Commonwealth* (1879). Kansas Historical Society Pamphlet Collection.

Peaslee, John B. "Arbor Day or Tree-Planting Celebration." *Proceedings of the American Forestry Congress at Its Meeting Held in Boston, September, 1885.* Washington, DC: Judd and Detweiler, 1886.

——. "Trees and Tree Planting, with Exercises and Directions for the Celebration of Arbor Day." In *Planting Trees in School Grounds and the Celebration of Arbor Day.* Department of the Interior, Bureau of Education. Washington, DC: GPO, 1885.

Periam, Jonathan. "Forest Tree Planting, as a Means of Wealth." *Transactions of the Illinois State Horticultural Society for 1871.* Chicago: Reade, Brewster, 1872.

Possibilities of Shelterbelt Planting in the Plains Region. Prepared under the direction of the Lake States Forest Experiment Station, USFS. Washington, DC: GPO, 1935.

Powell, John Wesley. "Our Recent Floods." *North American Review* 155, no. 429 (August 1892).

——. *Report on the Lands of the Arid Region of the United States.* 1878. Reprint, Cambridge, MA: Belknap Press, 1962.

——. "Trees on Arid Lands." *Science* 12, no. 297 (October 12, 1888).

Powers, Edward. "Artificial Rain." *American Meteorological Journal* 9, no. 4 (August 1892).

——. *War and the Weather.* 1871. Reprint, Delevan, WI: E. Powers, 1890.

Price, H. C. "Forestry and Its Effect on Western Climate." *Proceedings of the Iowa Park and Forestry Association. Second Annual Meeting, Des Moines, Iowa, December 8, 9, 10, 1902.* Iowa City, IA: Iowa Park and Forestry Association, 1903.

Rafter, George W. "The Relation of Rainfall to Run-Off." Water Supply and Irrigation Paper 80. United States Geological Survey, Department of the Interior. Washington, DC: GPO, 1903.

"Rainfall-Chart of the United States Showing the Distribution by Isohyetal Curves of the Mean Precipitation in Rain and Melted Snow for the Year—mean annual precipitation shown by isohyetal curves for every sixth inch from 8 to 68 inches." 2nd ed. Smithsonian Institution, Prof. J. Henry, secretary, including records to 1877. Prof. Spencer F. Baird, Secretary. Constructed by Charles A. Schott, Assistant USC and G Survey. Washington, DC, 1880.

"The Rain-Makers." *Monthly Weather Review of the Iowa Weather and Crop Service* 2, no. 8 (August 1891).

"Rain Production of Frank Melbourne during the Season of 1891." Nebraska State Historical Society Pamphlet 551.57 M49r.

Read, M. C. "The Preservation of Forests on the Headwaters of Streams" and "The Proper Value and Management of Government Timber Lands and the Distribution of North American Forest Trees." Papers read at the United States Department of Agriculture, May 7–8, 1884. Department of Agriculture, Miscellaneous, Special Report 5. Washington, DC: GPO, 1884.

Remy, Jules, and Julius Brenchley. *A Journey to Great-Salt-Lake-City.* London: W. Jeffs, 1861.

Report of the Public Lands Commission Created by the Act of March 3, 1879, Relating to the Public Lands in the Western Portion of the United States and to the Operation of Existing Land Laws. Washington, DC: GPO, 1880.

Roberts, Thomas P. "Relation of Forests to Floods." *Proceedings of the American Forestry Congress at Its Meeting Held in Boston, September 1885*. Washington, DC: Judd and Detweiler, 1886.

Robinson, Sara T. L. *Kansas: Its Interior and Exterior Life, Including a Full View of Its Settlement, Political History, Social Life, Climate, Soil, Productions, Scenery Etc.* Boston: Crosby, Nichols, 1856.

Sage, J. R. "Influence of Forests on Climate in Iowa." *American Meteorological Journal* 10, no. 14 (March 1894).

———. "The Practical Value of Reliable Crop and Weather Reports." *Monthly Review of the Iowa Weather and Crop Service, Co-operating with the United States Signal Service* 1, no. 8 (November 1890). Des Moines, IA: R. H. Ragsdale. State Historical Society of Iowa, Iowa City—Call No. QC.984.I6.

Schott, Charles A. "Rain Chart of the United States Showing by Isohyetal Lines the Distribution of the Mean Annual Precipitation in Rain and Melted Snow." 1868.

Sinks, Tiffin. "Report on the Climatology of Kansas" (1865). University of Kansas, Spencer Library, RH C5773.

Sixth Annual Meeting of the American Forestry Congress. Held in Springfield, IL, 1887. Springfield, IL: State Register Book and Job Print, 1887. Forest History Society Library, George H. Wirt Collection.

Snow, Frank H. "Change in the Climate of Kansas." *Transactions of the Kansas Academy of Science* 20, part 2. Topeka, KS: State Printing Office, 1907.

———. "Climate of Kansas." Report Submitted to Alfred Gray, Secretary State Agricultural Society, January 1st, 1873.

———. "Is the Rainfall of Kansas Increasing?" *Kansas City Review of Science and Industry*, vol. 8. Kansas City, MO: Ramsey, Millet, and Hudson, 1885.

———. "Periodicity of Kansas Rainfall and Possibilities of Storage of Excess Rainfall." *Ninth Biennial Report of the Kansas State Board of Agriculture*. Topeka, KS: Edwin H. Snow, 1895.

Stephens, E. F. "What Has the Timber Claim Law Done for Nebraska?" *Annual Report of the Nebraska State Horticultural Society for the Year 1897 Containing the Proceedings of the Annual Meeting Held at Lincoln, January, 1897*. Lincoln: State of Nebraska, 1897.

Stretch, Richard H., and J. E. James. "Reports on the Practicability of Turning the Waters of the Gulf of California into the Colorado Deserts and the Death Valley." S. Doc., 43rd Cong., 1st Sess., Misc. Doc. no. 84 (March 19, 1874).

Swain, George F. "The Influence of Forests upon the Rainfall and upon the Flow of Forests." *American Meteorological Journal* 7, no. 5 (November 1888).

Tewksbury, G. E. *The Kansas Picture Book*. Topeka, KS: A. S. Johnson, 1883.

Thompson, S. R. "The Rainfall of Nebraska." *American Meteorological Journal* 2, no. 1 (June 1884).

Tice, J. H. "Meteorological Effects of Forests." *Transactions of the Illinois State Horticultural Society for 1870*, Held at Galesburg, December 13, 14, 15, and 16. Chicago: Dunlop, Reade, and Brewster, 1871.

Topeka Commonwealth. "One Hundredth Meridian," 1879—Kansas Historical Society, Pamphlet Collection.

Transactions of the Illinois State Horticultural Society for 1871. Chicago: Reade, Brewster, 1872.

Transactions of the Iowa State Horticultural Society for 1879. Des Moines, IA: F. M. Mills, 1880.

Trowbridge, John. "Great Fires and Rain-Storms." *Popular Science Monthly*, December 1872.

Utah Board of Trade. *Resources and Attractions of the Territory of Utah*. Omaha, NE: Omaha Republican Publishing House, 1879.

Vennor, Henry G. *Vennor's Almanac and Weather Record for 1878–1879*. Montreal: Witness Printing House, 1879.

Warder, John A. [Member of the Scientific Commission of the United States—Vienna International Exposition, 1873]. Address Delivered before the Otoe County Horticultural Society in Nebraska City, September 12, 1878, "The Future Orchards and Forests of Nebraska." Nebraska State Historical Society, 634.9 W21a.

———. "Larch Wood." In *The American Journal of Forestry*, edited by Franklin B. Hough, vol. 1 (September 1882-October 1883). Cincinnati, OH: Robert Clarke, 1882–83.

———. *Report on Forests and Forestry*. Washington, DC: GPO, 1875.

———. "Some Trees for Planting on the Open Prairies of Northern Illinois and Adjoining Regions." *Transactions of the Illinois State Horticultural Society*, 1881.

———. "Tree Planting in Shelter Belts." *Journal of the American Agricultural Association*, May 1882.

———. "What Are Forest Trees?" Printed at the request of the secretary of the Kansas Horticultural Society and submitted to the commissioner of the General Land Office for "the purpose of securing a reconsideration of the ruling of the Department which excluded the Catalpa, Osage Orange, and Ailantus from being planted under the timber acts of Congress." North Bend, OH, January 10, 1882.

Watson, Winslow C. "Forests—Their Influence, Uses, and Reproduction." *Transactions of the New York State Agricultural Society for the Year 1865*. Albany, NY: Cornelius Wendell, 1866.

Welch, Rodney. "How the West Has Moved On." Address Delivered at Lincoln, September 27, 1877, during the Nebraska State Fair. Nebraska State Board of Agriculture, 1877. Nebraska State Historical Society, J. Sterling Morton Pamphlet Collection, vol. 4, no. 99.

Whitney, J. D. "Plain, Prairie, Forest." *American Naturalist* 10, no. 10 (October 1876).

Wilber, Charles Dana. *The Great Valleys and Prairies of Nebraska and the Northwest*. Omaha, NE: Daily Republican Printing, 1881.

Woiekoff. A. "Causes of Rainfall and Surface Conditions." *American Meteorological Journal* 10, no, 9 (January 1894).

Yingling, W. A. *Westward, or Central-Western Kansas*. Ness City, KS: Star Printing, 1890.

Zon, Raphael. *Forests and Water in Light of Scientific Investigation*. Washington, DC: GPO, 1927.

———. "How the Forests Feed the Clouds." In *Science Remaking the World*. New York: Doubleday, Page, 1923.

Secondary Sources

Adams, David K., et al. "The Amazon Dense GNSS Meteorological Network: A New Approach for Examining Water Vapor and Deep Convection in the Tropics." *Bulletin of the American Meteorological Society* 96 (December 2015).

Anderson, Benedict. *Imagined Communities: Reflections on the Origin and Spread of Nationalism*. Rev. ed. London: Verso, 2006.

Anderson, Katharine. "Mapping Meteorology." In *Intimate Universality: Local and*

Global Themes in the History of Weather and Climate, edited by James Rodger Fleming, Vladimir Jankovic, and Deborah R. Coen. Sagamore Beach, MA: Science History Publications, 2006.

Armitage, Kevin C. *The Nature Study Movement: The Forgotten Popularizer of America's
Conservation Ethic*. Lawrence: University of Kansas Press, 2009.

Baker, Zeke. "Agricultural Capitalism, Climatology and the 'Stabilization' of Climate in
the United States, 1850–1920." *British Journal of Sociology*, June 2020.

Baltensperger, B. H. "Plains Boomers and the Creation of the Great American Desert
Myth." *Journal of Historical Geography* 18 (1992).

Barnett, Cynthia. *Rain: A Cultural and Natural History*. New York: Crown, 2015.

Barnett, Lydia. "The Theology of Climate Change: Sin as Agency in the Enlightenment's
Anthropocene." *Environmental History* 20 (April 2015).

Barry, John M. *Rising Tide: The Great Mississippi Flood of 1927 and How It Changed
America*. New York: Simon and Schuster, 1997.

Batuman, Bülent. "The Shape of the Nation: Visual Production of Nationalism through
Maps in Turkey." *Political Geography* 29 (2010).

Beck, Ulrich, Anthony Giddens, and Scott Lash. *Reflexive Modernization: Politics, Tradition, and Aesthetics in the Modern Social Order*. Stanford, CA: Stanford University
Press, 1994.

Bederman, Gail. *Manliness and Civilization: A Cultural History of Gender and Race in the
United States, 1880–1917*. Chicago: University of Chicago Press, 1995.

Belyea, Barbara. "Inland Journeys, Native Maps." In *Cartographic Encounters: Perspectives on Native American Mapmaking and Map Use*, edited by Malcolm G. Lewis.
Chicago: University of Chicago Press, 1998.

Bergman, James. "Knowing Their Place: The Blue Hill Observatory and the Value of
Local Knowledge in an Era of Synoptic Weather Forecasting, 1884–1894." *Science in
Context* 29 (2016).

Berman, Marshall. *All That Is Solid Melts into Air: The Experience of Modernity*. New
York: Penguin Books, 1982.

Bernstein, David. *How the West Was Drawn: Mapping, Indians, and the Construction of
the Trans-Mississippi West*. Lincoln: University of Nebraska Press, 2018.

Bishop, Maude M. "Joseph Taplin Lovewell." *Bulletin of the Shawnee County Historical
Society* 38 (December 1962).

Bolster, Jeffrey. *The Mortal Sea: Fishing the Atlantic in the Age of Sail*. Cambridge, MA:
Harvard University Press, 2012.

Bradford, Marlene. "Historical Roots of Modern Tornado Forecasts and Warnings."
Weather and Forecasting 14 (August 1999).

Brock, Emily. "The Challenge of Reforestation: Ecological Experiments in the Douglas
Fir Forest." *Environmental History* 9 (2004).

Brückner, Martin. *The Geographic Revolution in Early America: Maps, Literacy, and National Identity*. Chapel Hill: University of North Carolina Press, 2006.

Buck, Holly Jean. "Geoengineering: Re-making Climate for Profit or Humanitarian
Intervention?" *Development and Change* 43 (2012).

Burge, Daniel. "Manifest Mirth: The Humorous Critique of Manifest Destiny, 1846–
1858." *Western Historical Quarterly* 47 (Autumn 2016): 283–302.

Cahan, David. "Looking at Nineteenth-Century Science: An Introduction." In *From
Natural Philosophy to the Sciences: Writing the History of Nineteenth-Century Science*,
edited by David Cahan. Chicago: University of Chicago Press, 2003.

Callon, Michel. "Some Elements of a Sociology of Translation: Domestication of Scallops and the Fishermen of St. Brieuc Bay." In *Power, Action, and Belief: A New Sociology of Knowledge?* Boston: Routledge, 1986.

Campbell, Brian L. "Uncertainty as Symbolic Action in Disputes among Experts." *Social Studies of Science* 15 (August 1985).

Canadell, J. G., and M. R. Raupach. "Managing Forests for Climate Change Mitigation." *Science* (AAAS) 320 (June 13, 2008).

Carey, Mark. "Science, Models, and Historians: Toward a Critical Climate History." *Environmental History* 19 (April 2014).

Carson, Sarah. "Atmospheric Happening and Weather Reasoning: Climate History in South Asia." *History Compass* 18 (2020).

Coen, Deborah. "Climate and Circulation in Imperial Austria." *Journal of Modern History* 82 (December 2010).

———. *Climate in Motion: Science, Empire, and the Problem of Scale.* Chicago: University of Chicago Press, 2018.

———. "Imperial Climatographies from Tyrol to Turkestan." *Osiris* 26 (2011).

———. *Vienna in the Age of Uncertainty: Science, Liberalism and Private Life.* Chicago: University of Chicago Press, 2007.

Colavito, Jason. *The Mound Builder Myth: Fake History and the Hunt for a "Lost White Race."* Norman: University of Oklahoma Press, 2020.

Conn, Steven. *History's Shadow: Native Americans and Historical Consciousness in the Nineteenth Century.* Chicago: University of Chicago Press, 2004.

Corfidi, Stephen F., Michael C. Coniglio, Ariel E. Cohen, and Corey M. Mead. "A Proposed Revision to the Definition of 'Derecho.'" *Bulletin of the American Meteorological Society* 97, no. 6 (June 2016).

Courtwright, Julie. "On the Edge of the Possible: Artificial Rainmaking and the Extension of Hope on the Great Plains." *Agricultural History* 89 (2015).

Craib, Raymond. *Cartographic Mexico: A History of State Fixations and Fugitive Landscapes.* Durham, NC: Duke University Press, 2004.

Crane, Gregg. "Playing It Safe: American Literature and the Taming of Chance." *Modern Intellectual History* 11 (2014).

Crawford, Robert P. *These Fifty Years: A History of the College of Agriculture.* Lincoln: University of Nebraska College of Agriculture, 1925.

Creech, Joe. "*The Tolerant Populists* and the Legacy of Walter Nugent." *Journal of the Gilded Age and Progressive Era* 14 (2015).

Cronon, William. *Nature's Metropolis: Chicago and the Great West.* New York: W. W. Norton, 1991.

Cullon, Joseph. "Legacies and Limitations: Environmental Historians Reconsider Progressive Conservation." *Journal of the Gilded Age and Progressive Era* (April 2002).

Culver, Lawrence. "Seeing Climate through Culture." *Environmental History* 19 (April 2014).

Cushman, Gregory. *Guano and the Opening of the Pacific World: A Global Ecological History.* Cambridge: Cambridge University Press, 2013.

Daipha, Phaedra. *Masters of Uncertainty: Weather Forecasters and the Quest for Ground Truth.* Chicago: University of Chicago Press, 2015.

———. "Weathering Risk: Uncertainty, Weather Forecasting, and Expertise." *Sociology Compass* 6 (2012).

Daston, Lorraine, and Peter Galison. *Objectivity.* New York: Zone Books, 2010.

Davis, Diana K. *The Arid Lands: History, Power, Knowledge.* Cambridge, MA: MIT
     Press, 2016.
————. *Resurrecting the Granary of Rome: Environmental History and French Colonial
     Expansion in North Africa.* Athens: Ohio University Press, 2007.
De Bres, Karen. "Come to the Champagne Air: Changing Promotional Images of the
     Kansas Climate, 1854–1900." *Great Plains Quarterly* 23 (Spring 2003).
Dodds, Gordon B. "The Stream-Flow Controversy: A Conservation Turning Point."
     Journal of American History 56 (June 1969).
Drake, Brian Allen. "Waving 'a Bough of Challenge': Forestry on the Kansas Grass-
     lands." *Great Plains Quarterly* 23 (Winter 2003).
Drake, James D. *The Nation's Nature: How Continental Presumptions Gave Rise to the
     United States of America.* Charlottesville: University of Virginia Press, 2011.
Droze, Wilmon H. *A History of Tree Planting in the Plains States.* Denton: Texas Woman's
     University, 1977.
Edwards, Paul. "Meteorology as Infrastructural Globalism." *Osiris* 21 (2006).
————. *A Vast Machine: Computer Models, Climate Data, and the Politics of Global
     Warming.* Cambridge, MA: MIT Press, 2010.
Edwards, Rebecca. *New Spirits: Americans in the Gilded Age, 1865–1905.* Oxford: Oxford
     University Press, 2006.
Emmons, David. *Garden in the Grasslands: Boomer Literature of the Central Great Plains.*
     Lincoln: University of Nebraska Press, 1971.
————. "Theories of Increased Rainfall and the Timber Culture Act of 1873." *Forest His-
     tory* 15 (October 1971).
Farmer, Jared. *On Zion's Mount: Mormons, Indians, and the American Landscape.* Cam-
     bridge, MA: Harvard University Press, 2008.
————. *Trees in Paradise: A California History.* New York: W. W. Norton, 2013.
Faust, Drew Gilpin. *This Republic of Suffering: Death and the American Civil War.* New
     York: Vintage, 2008.
Fiege, Mark. *Irrigated Eden: The Making of an Agricultural Landscape in the American
     West.* Seattle: University of Washington Press, 1999.
Fink, Leon. *The Long Gilded Age: American Capitalism and the Lessons of a New World
     Order.* Philadelphia: University of Pennsylvania Press, 2015.
Fite, Gilbert. *The Farmer's Frontier, 1865–1900.* New York: Holt, Rinehart, and Winston,
     1966.
Flanagan, Maureen A. *America Reformed: Progressives and Progressivisms, 1890s-1920s.*
     New York: Oxford University Press, 2007.
Fleming, James Rodger. "Climate, Change, History." *Environment and History* 20 (2014).
————. *Fixing the Sky: The Long and Checkered History of Weather and Climate Control.*
     New York: Columbia University Press, 2010.
————. *Historical Perspectives on Climate Change.* Oxford: Oxford University Press,
     1998.
————. *Meteorology in America, 1800–1870.* Baltimore: Johns Hopkins University Press,
     1990.
Foster, Mike. *Strange Genius: The Life of Ferdinand Vandeveer Hayden.* Niwot, CO: Rob-
     erts Rinehard, 1994.
Francaviglia, Richard. *The Mapmakers of New Zion: A Cartographic History of Mormon-
     ism.* Salt Lake City: University of Utah Press, 2015.
————. *The Mormon Landscape: Existence, Creation, and Perception of a Unique Image of
     the American West.* New York: AMS Press, 1978.

Freeman, John F. *High Plains Horticulture: A History*. Boulder: University of Colorado Press, 2008.

Fritzsche, Peter. *Reading Berlin*. Cambridge, MA: Harvard University Press, 1996.

Gardner, Robert. "Constructing a Technological Forest: Nature, Culture, and Tree-Planting in the Nebraska Sand Hills." *Environmental History* 14 (April 2009).

Gieryn, Thomas. *Cultural Boundaries of Science: Credibility on the Line*. Chicago: University of Chicago Press, 1999.

Golding, Michael, Helen Tilley, and Gyan Prakash. *Utopia/Dystopia: Conditions of Historical Possibility*. Princeton, NJ: Princeton University Press, 2010.

Grove, Richard. *Green Imperialism: Colonial Expansion, Tropical Island Edens and the Origins of Environmentalism*. Cambridge: Cambridge University Press, 1995.

Guha, Ramachandra. *Environmentalism: A Global History*. New York: Longman, 2000.

Hacking, Ian. *The Taming of Chance*. Cambridge: Cambridge University Press, 1990.

Hall, Marcus. *Earth Repair: A Transatlantic History of Environmental Restoration*. Charlottesville: University of Virginia Press, 2005.

Hamerla, Ralph R. *An American Scientist on the Research Frontier: Edward Morley, Community, and Radical Ideas in Nineteenth-Century Science*. Dordrecht, Netherlands: Springer, 2006.

Hargreaves, Mary Wilma. *Dry Farming in the Northern Great Plains, 1900–1925*. Cambridge, MA: Harvard University Press, 1957.

Harley, J. B. "Texts and Contexts in the Interpretation of Early Maps." In *The New Nature of Maps; Essays in the History of Cartography*, edited by Paul Laxton. Baltimore: Johns Hopkins University Press, 2001.

Harper, Kristine C. *Make It Rain: State Control of the Atmosphere in Twentieth-Century America*. Chicago: University of Chicago Press, 2017.

Hays, Samuel P. *Conservation and the Gospel of Efficiency: The Progressive Conservation Movement, 1890–1920*. 1959. Reprint, New York: Atheneum, 1972.

Heymann, Matthias. "The Climate Change Dilemma: Big Science, the Globalizing of Climate and the Loss of the Human Scale." *Regional Environmental Change* 19 (2019).

Hoganson, Kristin. *Fighting for American Manhood: How Gender Politics Provoked the Spanish-American and Philippine-American Wars*. New Haven, CT: Yale University Press, 1998.

Howe, Joshua P. *Behind the Curve: Science and the Politics of Global Warming*. Seattle: University of Washington Press. 2014.

Hulme, Mike. "Climate and Its Changes: A Cultural Appraisal." *Geo* 2 (2015).

———. *Why We Disagree about Climate Change*. Cambridge: Cambridge University Press, 2009.

Ingold, Tim. *Perception of the Environment: Essays on Livelihood, Dwelling, and Skill*. New York: Routledge, 2000.

Jacob, Nancy. *Birders of Africa: History of a Network*. New Haven, CT: Yale University Press, 2016.

Jacoby, Karl. *Crimes against Nature: Squatters, Poachers, Thieves, and the Hidden History of American Conservation*. Berkeley: University of California Press, 2003.

———. *Shadows at Dawn: An Apache Massacre and the Violence of History*. New York: Penguin, 2009.

Jankovic, Vladimir, and Andrew Bowman. "After the Green Gold Rush: The Construction of Climate Change as a Market Transition." *Economy and Society* 43 (2014).

Johnson, Benjamin H. *Escaping the Dark, Gray City: Fear and Hope in Progressive-Era Conservation*. New Haven, CT: Yale University Press, 2017.

Johnston, Robert. *The Radical Middle Class: Populist Democracy and the Question of Capitalism in Progressive Era Portland, Oregon.* Princeton, NJ: Princeton University Press, 2003.

Kern, Stephen. *The Culture of Time and Space, 1880–1918.* Cambridge, MA: Harvard University Press, 1983.

Kiser, William S. *Coast-to-Coast Empire: Manifest Destiny and the New Mexico Borderlands.* Norman: University of Oklahoma Press, 2018.

Kivelson, Valerie. *Cartographies of Tsardom: the Land and Its Meanings in Seventeenth-Century Russia.* Ithaca, NY: Cornell University Press, 2006.

Klein, Naomi. *This Changes Everything: Capitalism vs. the Climate.* New York: Simon and Schuster, 2014.

Kline, Ronald. "Construing 'Technology' as 'Applied Science': Public Rhetoric of Scientists and Engineers in the United States, 1880–1945." *Isis* 86 (1995).

Klingle, Matthew. *Emerald City: An Environmental History of Seattle.* New Haven, CT: Yale University Press, 2007.

Kohout, Amy. "From the Field: Nature and Work on American Frontiers, 1876–1909." PhD diss., Cornell University, 2015.

Kollmorgen, Johanna, and Walter Kollmorgen. "Landscape Meteorology in the Plains Area." *Annals of the Association of American Geographers* 63 (December 1973).

Kollmorgen, Walter. "The Woodsman's Assault on the Domain of the Cattleman." *Annals of the Association of American Geographers* 59, no. 2 (1969).

Kral, E. A. "Charles Dana Wilber: Scientific Promoter, Pioneer of the West and Town Founder." *Wilber Republican*, August 2, 2000.

Krech, Shepard. *The Ecological Indian: Myth and History.* New York: W. W. Norton, 1999.

Kroeker, Marvin E. "Deceit about the Garden: Hazen, Custer, and the Arid Lands Controversy." *North Dakota Quarterly* 38 (Summer 1970).

———. *Great Plains Command: William B. Hazen in the Frontier West.* Norman: University of Oklahoma Press, 1976.

Krupar, Shiloh R. *Hot Spotter's Report: Military Fables of Toxic Waste.* Minneapolis: University of Minnesota Press, 2013.

Kutzleb, Charles. "Can Forests Bring Rain to the Plains?" *Forest History* 15 (October 1971).

———. "Rain Follows the Plow: History of an Idea." PhD diss., University of Colorado, 1968.

Langston, Nancy. *Forest Dreams, Forest Nightmares: The Paradox of Old Growth in the Inland West.* Seattle: University of Washington Press, 1995.

Lasch, Christopher. *The True and Only Heaven: Progress and Its Critics.* New York: Norton, 1991.

Latour, Bruno. *The Politics of Nature: How to Bring the Sciences into Democracy.* Translated by Catherine Porter. Cambridge, MA: Harvard University Press, 2004.

———. *Science in Action: How to Follow Scientists and Engineers through Society.* Cambridge, MA: Harvard University Press, 1987.

———. *We Have Never Been Modern.* Cambridge, MA: Harvard University Press, 1993.

Lears, T. J. Jackson. *No Place of Grace: Antimodernism and the Transformation of American Thought, 1880–1920.* New York: Pantheon, 1981.

———. *Rebirth of a Nation: The Making of Modern America.* New York: HarperCollins, 2009.

————. *Something for Nothing: Luck in America*. New York: Viking, 2003.

Legg, Stephen. "Debating the Climatological Role of Forests in Colonial Victoria and South Australia." In *Climate, Science, and Colonization: Histories from Australia and New Zealand*, edited by James Beattie, Emily O'Gorman, and Matthew Henry. New York: Palgrave Macmillan, 2014.

Lehmann, Philipp. "Infinite Power to Save the World: Hydroelectricity and Engineered Climate Change in the Alantropa Project." *American Historical Review* 121 (2016).

————. "Whither Climatology? Brückner's *Climate Oscillations*, Data Debates, and Dynamic Climatology." *History of Meteorology* 7 (2015).

LeMenager, Stephanie. *Manifest and Other Destinies: Territorial Fictions of the Nineteenth Century*. Lincoln: University of Nebraska Press, 2004.

Levy, Jonathan. *Freaks of Fortune: The Emerging World of Capitalism and Risk in America*. Cambridge, MA: Harvard University Press, 2012.

Libecap Gary D., and Zeynep K. Hansen. "Rain Follows the Plow and Dryfarming Doctrine: The Climate Information Problem and Homestead Failure in the Upper Great Plains." *Journal of Economic History* 62 (March 2002).

Lightman, Bernard. *Victorian Popularizers of Knowledge: Designing Nature for New Audiences*. Chicago: University of Chicago Press, 2007.

Limerick, Patricia Nelson. *The Legacy of Conquest: The Unbroken Past of the American West*. New York: Norton, 1987.

Lingenfelter, Richard E. *Death Valley and the Amargosa: A Land of Illusion*. Berkeley: University of California Press, 1986.

Locher, Fabien, and Jean-Baptiste Fressoz. "Modernity's Frail Climate: A Climate History of Environmental Reflexivity." *Critical Inquiry* 38 (Spring 2012).

Lowenthal, David. *George Perkins Marsh: Prophet of Conservation*. Seattle: University of Washington Press, 2000.

Lucier, Paul. "The Origins of Pure and Applied Science in Gilded Age America." *Isis* 103 (2012).

————. "The Professional and the Scientist in Nineteenth-Century America." *Isis* 100 (2009).

Maggor, Noam. *Brahmin Capitalism: Frontiers of Wealth and Populism in America's First Gilded Age*. Cambridge, MA: Harvard University Press, 2017.

Maher, Neil. *Nature's New Deal: The Civilian Conservation Corps and the Roots of the American Conservation Movement*. Oxford: Oxford University Press, 2008.

McGerr, Michael. *A Fierce Discontent: The Rise and Fall of the Progressive Movement in the United States, 1870–1920*. New York: Free Press, 2003.

McIntosh, C. Barron. "The Use and Abuse of the Timber Culture Act." *Annals of the Association of American Geographers* 65 (September 1975).

Mergen, Bernard. *Weather Matters: An American Cultural History since 1900*. Lawrence: University of Kansas Press, 2008.

Meyer, William. *Americans and Their Weather*. Oxford: Oxford University Press, 2000.

Miller, Char. *Gifford Pinchot and the Making of Modern Environmentalism*. Washington, DC: Island Press, Shearwater Books, 2001.

Miller, Daegan. "Reading Tree in Nature's Nation: Toward a Field Guide to Sylvan Literacy in the Nineteenth-Century United States." *American Historical Review* 121 (October 2016).

Miner, Craig. *West of Wichita: Settling the High Plains of Kansas, 1865–1890*. Lawrence: University of Kansas Press, 1986.

Mitchell, Timothy. *Rule of Experts: Egypt, Techno-Politics, Modernity*. Berkeley: University of California Press, 2002.

Mitman, Gregg. "Geographies of Hope: Mining the Frontiers of Health in Denver and Beyond." *Osiris* 19 (2004).

Monmonier, Mark. *Cartographies of Danger: Mapping Hazards in America*. Chicago: University of Chicago Press, 1997.

Montoya, Maria. *Translating Property: The Maxwell Land Grant and the Conflict over Land in the American West*. Berkeley: University of California Press, 2002.

Moon, David. *The American Steppes: The Unexpected Russian Roots of Great Plains Agriculture, 1870s-1930s*. Cambridge: Cambridge University Press, 2020.

———. *The Plow That Broke the Steppes: Agriculture and Environment on Russia's Grasslands, 1700–1914*. Oxford: Oxford University Press, 2013.

Murphy, Michelle. *Sick Building Syndrome and the Problem of Uncertainty*. Durham, NC: Duke University Press, 2006.

Nash, Linda. *Inescapable Ecologies: A History of Environment, Disease, and Knowledge*. Berkeley: University of California Press, 2006.

Naylor, Simon. "Nationalizing Provincializing Weather: Meteorology in Nineteenth-Century Corwall." *British Journal for the History of Science* 39 (2006).

Nobles, Gregory H. "Straight Lines and Stability: Mapping the Political Order of the Anglo-American Frontier." *Journal of American History* 80 (June 1993)

Nye, David. *America as Second Creation: Technology and Narratives of New Beginnings*. Cambridge, MA: MIT Press, 2003.

Olsen, James C. "Arbor Day—A Pioneer Expression of Concern for Environment." *Nebraska History* 53 (1972).

Oreskes, Naomi, and Erik Conway. *Merchants of Doubt: How a Handful of Scientists Obscured the Truth on Issues from Tobacco Smoke to Global Warming*. New York: Bloomsbury Press, 2010.

Orth, Joel J. "The Conservation Landscape: Trees and Nature on the Great Plains." PhD diss., Iowa State University, 2004.

———. "Directing Nature's Creative Forces: Climate Change, Afforestation, and the Nebraska National Forest." *Western Historical Quarterly* 42 (Summer 2011).

———. "The Shelterbelt Project: Cooperative Conservation in 1930s America." *Agricultural History* 81 (Summer 2007).

Palmer, W. P. "Dissent at the University of Iowa: Gustavus Detlef Hinrichs—Chemist and Polymath." *Chemistry* 16, no. 6 (2007).

Pandora, Katherine. "Knowledge Held in Common: Tales of Luther Burbank and Science in the American Vernacular." *Isis* 92 (September 2001).

Pauly, Philip. *Fruits and Plains: The Horticultural Transformation of America*. Cambridge, MA: Harvard University Press, 2007.

Peluso, Nancy Lee. "Whose Woods Are These? Counter-Mapping Forest Territories in Kalimantan, Indonesia." In *The Anthropology of Development and Globalization*, edited by Marc Edelman and Angelique Haugerud. Malden, MA: Blackwell, 2005.

Perkins, Sid. "Crop Irrigation Could Be Cooling Midwest." *Science News*, January 22, 2010.

Pietruska, Jamie L. "Hurricanes, Crops, and Capital: The Meteorological Infrastructure of American Empire in the West Indies." *Journal of the Gilded Age and Progressive Era* 15 (October 2016).

———. *Looking Forward: Prediction and Uncertainty in Modern America*. Chicago: University of Chicago Press, 2017.

————. "US Weather Bureau Chief Willis Moore and the Reimagination of Uncertainty in Long-Range Forecasting." *Environment and History* 17 (2011).

Pinch, Trevor. "The Sun-Set: The Presentation of Certainty in Scientific Life." *Social Studies of Science* 11 (February 1981).

Pisani, Donald J. "Forests and Conservation, 1865–1890." In *American Forests: Nature, Culture, and Politics*, edited by Char Miller. Lawrence: University of Kansas Press, 1997.

————. *Water and American Government: The Reclamation Bureau, National Water Policy, and the West, 1902–1935*. Berkeley: University of California Press, 2002.

————. *Water, Land, and Law in the West: The Limits of Public Policy, 1850–1920*. Lawrence: University of Kansas Press, 1996.

Pritchard, Sara B. "Joining Environmental History with Science and Technology Studies: Promises, Challenges, and Contributions." In *New Natures: Joining Environmental History with Science and Technology Studies*, edited by Sara B. Pritchard, Finn Arne Jorgensen, and Dolly Jorgensen. Pittsburgh: University of Pittsburgh Press, 2013.

Proctor, Robert N. *Cancer Wars*. New York: Basic Books, 1995.

Quinn, Katrine. "'Across the Continent . . . and Still the Republic!' Inscribing Nationhood in Samuel Bowles's Newspaper Letters of 1865." *American Journalism* 31, no. 4 (2014).

Riley, John L. *The Once and Future Great Lakes Country: An Ecological History*. Montreal: McGill University Press, 2013.

Rodgers, Andrew Denny. *Bernhard Eduard Fernow: A Story of North American Forestry*. Durham, NC: Forest History Society, 1991.

Rozario, Kevin. *The Culture of Calamity: Disaster and the Making of Modern America*. Chicago: University of Chicago Press, 2007.

Rutkow, Eric. *American Canopy: Tress, Forests, and the Making of a Nation*. New York: Scribner, 2012.

Sachs, Aaron. "American Arcadia: Mount Auburn Cemetery and the Nineteenth-Century Landscape Tradition." *Environmental History* 15 (April 2010).

————. *Arcadian America: The Death and Life of an Environmental Tradition*. New Haven, CT: Yale University Press, 2013.

————. *The Humboldt Current: Nineteenth Century Exploration and the Roots of American Environmentalism*. New York: Penguin, 2006.

Sackman, Douglas. *Wild Men: Ishi and Kroeber in the Wilderness of Modern America*. Oxford: Oxford University Press, 2010.

Sandage, Scott. *Born Losers: A History of Failure in America*. Cambridge, MA: Harvard University Press, 2005.

Schiebinger, Londa. "Agnotology and Exotic Abortifacients: The Cultural Production of Ignorance in the Eighteenth-Century World." *Proceedings of the American Philosophical Society* 149 (September 2005).

Schivelbusch, Wolfgang. *The Railway Journey: The Industrialization of Space and Time in the 19th Century*. Berkeley: University of California Press, 1986.

Schulten, Susan. *Mapping the Nation: History and Cartography in Nineteenth-Century America*. Chicago: University of Chicago Press, 2012.

Scott, James. *Seeing Like a State: How Certain Schemes to Improve the Human Condition Have Failed*. New Haven, CT: Yale University Press, 1998.

Scott, Joan. "The Evidence of Experience." *Critical Inquiry* 17 (1991).

Singerman, David. "Science, Commodities, and Corruption in the Gilded Age." *Journal of the Gilded Age and Progressive Era* 15 (2016).

Smiley, David L. *Lion of Whitehall: The Life of Cassius M. Clay.* Madison: University of Wisconsin Press, 1961.

Smith, Henry Nash. "Rain Follows the Plow: The Notion of Increased Rainfall for the Great Plains, 1844–1880." *Huntington Library Quarterly* 10 (February 1947).

Smith, Neil. *Uneven Development: Nature, Capital, and the Production of Space* (Oxford: Blackwell, 1984)

Specht, Joshua. "The Rise, Fall, and Rebirth of the Texas Longhorn: An Evolutionary History." *Environmental History* 21 (April 2016).

Spence, Mark. *Dispossessing the Wilderness: Indian Removal and the Making of the National Parks.* New York: Oxford University Press, 1999.

Star, Susan Leigh. "Scientific Work and Uncertainty." *Social Studies of Science* 15 (August 1985).

Steen, Harold K. *The US Forest Service: A History.* Seattle: University of Washington Press, 1976.

Stoll, Mark. "Religion 'Irradiates' the Wilderness." In *American Wilderness: A New History*, edited by Michael Lewis. Oxford: Oxford University Press, 2007.

Stoll, Steven. *Larding the Lean Earth: Soil and Society in Nineteenth-Century America.* New York: Hill and Wang, 2003.

Strauss, Sarah. "Weather Wise: Speaking Folklore to Science in Leukerbad." In *Weather, Climate, Culture*, edited by Saraj Strauss and Benjamin Orlove. New York: Berg, 2003.

Streck, C., and S. M. Scholz. "The Role of Forests in Global Climate Change: Whence We Come and Where We Go." *International Affairs* 82 (October 4, 2006).

Streeter, Floyd Benjamin. *The Kaw: The Heart of a Nation.* New York: Farrar and Rinehart, 1941.

Stroud, Ellen. *Nature Next Door: Cities and Trees in the American Northeast.* Seattle: University of Washington Press, 2012.

Sutter, Paul. "A Retreat from Profit: Colonization, the Appalachian Trail, and the Social Roots of Benton MacKaye's Wilderness Advocacy." *Environmental History* 4 (October 1999).

Swanberg, Susan E. "'The Way of the Rain': Towards a Conceptual Framework for the Retrospective Examination of Historical American and Australian 'Rain Follows the Plough/Plow' Messages." *International Review of Environmental History* 5 (2019).

Taylor, Dorceta E. *The Rise of the American Conservation Movement: Power, Privilege, and Environmental Protection.* Durham, NC: Duke University Press, 2016.

Temple, Samuel. "Forestation and Its Discontents: The Invention of an Uncertain Landscape in Southwestern France, 1850-Present." *Environment and History* 17 (2011).

Thompson, Kenneth. "Forests and Climate Change in America: Some Early Views." *Climatic Change* 3 (1980).

Thuesen, Peter J. *Tornado God: American Religion and Violent Weather.* New York: Oxford University Press, 2020.

Tilley, Helen. "Ecologies of Complexity: Tropical Environments, African Trypanosomiasis, and the Science of Disease Control in British Colonial India." *Osiris* 19 (2004).

———. "Global Histories, Vernacular Science, and African Genealogies, or Is the History of Science Ready for the World?" *Isis* 101 (March 2010).

Trachtenberg, Alan. *The Incorporation of America: Culture and Society in the Gilded Age.* New York: Hill and Wang, 1982.

Travis, Paul. "Changing Climate in Kansas: A Late 19th-Century Myth." *Kansas History* 1 (Spring 1978).

Truett, Samuel. *Fugitive Landscapes: The Forgotten History of the U.S.-Mexico Border-lands*. New Haven, CT: Yale University Press, 2006.

Tyrrell, Ian. *Crisis of the Wasteful Nation: Empire and Conservation in Theodore Roose-velt's America*. Chicago: University of Chicago Press, 2015.

Valenčius, Conevery Bolton. *The Health of the Country: How American Settlers Under-stood Themselves and Their Land*. New York: Basic Books, 2002.

———. *The Lost History of the New Madrid Earthquakes*. Chicago: University of Chi-cago Press, 2013.

Vetter, Jeremy. *Field Life: Science in the American West during the Railroad Era*. Pitts-burgh: University of Pittsburgh Press, 2016.

———. "Knowing the Great Plains Weather: Field Life and Lay Participation on the American Frontier during the Railroad Era." *East Asian Science, Technology, and So-ciety: An International Journal* 13 (2019).

———. "Lay Observers, Telegraph Lines, and Kansas Weather: The Field Network as a Mode of Knowledge Production." *Science in Context* 24 (June 2011).

Walker, Richard. *The Country in the City: The Greening of the San Francisco Bay Area*. Seattle: University of Washington Press, 2007.

Walter Prescott Webb, *The Great Plains*. 1931. Reprint, Waltham, MA: Blaisdell, 1959.

Warren, Louis. *Buffalo Bill's America*. New York: Knopf, 2005.

———. *The Hunter's Game: Poachers and Conservationists in Twentieth-Century Amer-ica*. New Haven, CT: Yale University Press, 1997.

West, Elliott. *The Contested Plains: Indians, Goldseekers and the Rush to Colorado*. Law-rence: University of Kansas Press, 1998.

Whitaker, Michael R. "Making War on Jupiter Pluvius: The Culture and Science of Rainmaking in the Southern Great Plains, 1870–1913." *Great Plains Quarterly* 33 (Fall 2013).

White, Richard. *Railroaded: The Transcontinentals and the Making of Modern America*. New York: Norton, 2011.

Wiersema, Courtney. "A Fruitful Plain: Fertility on the Shortgrass Prairie." *Environmen-tal History* 16 (2011).

Williams, Michael. *Americans and Their Forests: A History of Geography*. Cambridge: Cambridge University Press, 1989.

Wishart, David J. *The Last Days of the Rainbelt*. Lincoln: University of Nebraska Press, 2013.

Withers, Charles, and David Livingstone. "Thinking Geographically about Nineteenth-Century Science." In *Geographies of Nineteenth-Century Science*, edited by Charles Withers and David Livingstone. Chicago: University of Chicago Press, 2011.

Wood, Denis. "Mapmaking, Counter-Mapping, and Map Art in the Mapping of Pales-tine." In *Rethinking the Power of Maps*, edited by Denis Wood, John Fels, and John Krygier. New York: Guilford Press, 2010.

———. *The Power of Maps*. New York: Guilford Press, 1992.

Worster, Donald. *Dust Bowl: The Southern Plains in the 1930s*. Oxford: Oxford University Press, 1979.

———. *Nature's Economy: A History of Ecological Ideas*. 1977. Reprint, Cambridge: Cam-bridge University Press, 1985.

———. *Rivers of Empire: Water, Aridity, and the Growth of the American West*. New York: Pantheon Books, 1985.

Yearley, Steven. "The Environmental Challenge to Science Studies." In the *Handbook of Science and Technology Studies*. Thousand Oaks, CA: Sage Publications, 1995.

Zarefsky, David. *Lincoln, Douglas, and Slavery*. Chicago: University of Chicago Press, 1993.

Zilberstein, Anya. *A Temperate Empire: Making Climate Change in Early America*. New York: Oxford University Press, 2017.

Zizzamia, Daniel. "Restoring the Paleo-West: Fossils, Coal, and Climate in Late Nineteenth-Century America." *Environmental History* 24 (2019).

INDEX

Page numbers in italics refer to figures.